Exploring Complexity

Exploring Complexity

AN INTRODUCTION

Grégoire Nicolis
Ilya Prigogine

W. H. FREEMAN AND COMPANY / NEW YORK

Jacket cover image Courtesy of Stefan C. Müller, Theo Plesser, and Benno Hess, Max-Planck-Institut, Dortmund, Federal Republic of Germany

The spiral is a widespread structural element of patterns in nature. It also plays an important role in spatial self-organization during chemical modification of matter. The cover illustration shows the cores of several spiral waves formed during the oxidation of malonic acid in the presence of bromo-compounds (the Belousov-Zhabotinskii reaction). The three-dimensional perspective from "below" visualizes the envelopes of deep "concentration holes" around the rotation centers of the spirals. These concentration holes remain steady at their centers due to the interaction of diffusion and reaction. They demonstrate the singular character of the chemical processes in the core area and very closely resemble the shape of tornadoes.

Library of Congress Cataloging-in-Publication Data

Nicolis, G., 1939–
      Exploring complexity.

      Includes index.
        1. Science—Philosophy.   2. Complexity (Philosophy)
I. Prigogine, I. (Ilya)   II. Title.
Q175.N417 1989   501   88-33555
ISBN 0-7167-1859-6
ISBN 0-7167-1860-X (pbk.)

Printed in the United States of America

        4   5   6   7   8   9   VB   9   9   8   7   6   5   4

Contents

Preface

Our physical world is no longer symbolized by the stable and periodic planetary motions that are at the heart of classical mechanics. It is a world of instabilities and fluctuations, which are ultimately responsible for the amazing variety and richness of the forms and structures we see in nature around us. New concepts and new tools are clearly necessary to describe nature, in which evolution and pluralism become the key words. This book provides a short introduction to the methods devised over recent decades to explore *complexity*, be it at the level of molecules, of biological systems, or even of social systems.

We stress the role of two disciplines that have dramatically modified our outlook on complexity. The first is nonequilibrium physics. In this discipline the most unexpected outcome is the discovery of fundamental new properties of matter in far-from-equilibrium conditions. The second discipline is the modern theory of dynamical systems. Here, the central discovery is the prevalence of instability. Briefly, this means that small changes in initial conditions may lead to large amplifications of the effects of the changes.

The new methods developed in this context lead to a better understanding of the environment in which we live. In this environment we find both unexpected regularities as well as equally unexpected large-scale fluctuations. As evidence of regularity: matter is associated with an overwhelming dominance of particles over antiparticles, and life with a dominance of chiral, asymmetric biomolecules over their symmetrical opposites. What could have been the selection mechanism giving rise to such large-scale regularities? Conversely, we could have expected uniformity and stability of our climatic conditions. However, contrary to such expectations, climate has fluctuated violently over periods quite short as compared to the characteristic time of the evolution of the sun. How is this possible? We now begin to have methods to address these questions.

The first chapter of this book presents selected examples of complex phenomena arising in the framework of physico-chemical and biological systems, as well as in our environment at large. This description brings out a number of concepts that deal with mechanisms that are encountered repeatedly throughout the different phenomena; they are nonequilibrium, stability, bifurcation and symmetry breaking, and long-range order. In Chapter 2 these concepts are taken up and analyzed in more detail. They become the basic elements of what we believe to be a new scientific vocabulary, the *vocabulary of complexity*.

Following these two purely descriptive chapters, Chapter 3 addresses the problem of complexity from the standpoint of the modern theory of dynamical systems. We discuss some mechanisms by which nonlinear systems driven away from equilibrium can generate instabilities that lead to bifurcations and symmetry breaking. Special emphasis is placed in our analysis on the emergence of *chaotic dynamics*, the natural tendency of large classes of systems to evolve to states displaying both deterministic behavior and unpredictability.

Chapter 4 attempts a more detailed description of complex phenomena, going beyond the purely phenomenological level of the preceding chapters. We present the basic elements of probabilistic analysis of nonlinear nonequilibrium systems and construct a microscopic model of bifurcation and evolution. We also discuss some ways by which the concept of information can be integrated in the description of dynamical systems.

In the classical view, there was a sharp distinction between chance and necessity, between stochastic and deterministic behavior. The analysis in Chapters 3 and 4 shows that the situation is much more subtle. There are various forms of randomness, some of which are associated with the chaotic behavior of the solutions of simple deterministic equations. Chapter 5 addresses the question of the origin of randomness and irreversibility. We also discuss the closely related problem of understanding entropy and, in fact, the very concept of time. We believe that we begin to decipher the message of the celebrated second law of thermodynamics. We are living in a world of unstable processes, and this allows us to define an entropy function. Moreover, we live in a world in which the symmetry between past and future is broken; a world in which irreversible processes lead to equilibrium in our future. This universal prevalence of the breaking of time symmetry is at the heart of the second law.

We have expressed our conviction that science is bound to play an increasingly important role in our effort to understand our global environment. The ability to break the disciplinary barriers and to try new ways of looking at sometimes longstanding problems is therefore one of the essential goals of the methods of analysis of complex phenomena set forth in this book. Chapter 6 demonstrates how this transfer of knowledge from one field to another can be envisaged. We devote much of this final chapter to questions that are beyond the realm of tranditional concern in the physical sciences, such as the dynamics of climatic change, and the behavior of social insects and human populations. Obviously, each of these problems has its own specificities, and the possibility of a broad generalization should in no way be anticipated. Still, the role of nonlinearities and of fluctuations appears very clearly. It strongly suggests that the modeling of such systems should benefit from the new perspectives that the study of complex phenomena in nonlinear dynamical systems has provided to science.

Finally, Appendixes 1 to 5 are devoted to a more quantitative survey of some of the techniques used in the main portion of the book.

In preparing this book we have benefited greatly from discussions with numerous colleagues and coworkers. We are most grateful to E. Rebhan for a critical reading of the original manuscript, for suggesting numerous improvements, and for coordinating the translation into German. It is a pleasure to acknowledge the suggestions and help received from C. Baesens, F. Baras, J.L. Deneubourg, Y. Elskens, R. Feistel, H. Frisch, R. Mazo, M. Malek Mansour, C. Nicolis, J.S. Nicolis, G. Rao, and S. Rao.

Finally, it is also a pleasure to thank P. Pape, A. Kimmons, S. Wellens, and M. Adam for the difficult task of typing two successive versions of the manuscript, and P. Kinet for his efficient technical assistance.

The research of the authors in the field of irreversible phenomena and nonlinear dynamics is supported by the International Solvay Institutes of Physics and Chemistry, the Belgian Ministries of Education and Scientific Research, the Commission of the European Communities, the R.A. Welch Foundation (Houston, Texas), and the U.S. Department of Energy.

G. Nicolis
I. Prigogine

Exploring Complexity

Prologue: Science in an Age of Transition

Whatever our professional preoccupations may be, we cannot escape the feeling that we live in an age of transition, an age that demands constructive modification of our environment. We must find and explore new resources, must understand our environment better, and must achieve a less destructive coexistence with nature. The time scale of the qualitative modifications that are required to achieve these major goals is not comparable to the immense time spans involved in biological or geological evolution. Rather, it is of the order of the decade. Thus the modifications that must be made interfere with our own lives and the lives of the next generation.

We cannot anticipate the outcome of this period of transition, but it is clear that science is bound to play an increasingly important role in our effort to meet the challenge of understanding and reshaping our global environment. It is a striking fact that at this crucial moment science is going through a period of reconceptualization.

The two great revolutions in physics at the beginning of this century were quantum mechanics and relativity. Both started as corrections to classical mechanics, made necessary once the roles of the universal constants c (the velocity of light) and h (Planck's constant) were discovered. Today, both of these areas of physics have taken an unexpected "temporal" turn: Quantum mechanics now deals in its most interesting part with the description of unstable particles and their mutual transformations. Relativity, which started as a geometrical theory, today is mainly associated with the thermal history of the universe.

The concerns of these two areas—elementary particles and cosmology—correspond to extreme conditions; they are parts of high energy physics. On our own scale as well, the macroscopic scale of "everyday," real-world, directly observable events, physics is undergoing a radical transformation. Even a few years ago, a physicist asked about what is known and what is not known would have answered that the real problems were only at the frontiers of our universe, at the level of elementary particles and at the level of cosmology. In contrast, he or she would have asserted, the basic laws that are relevant on our macroscopic level were well known. Today, a growing minority of scientists would question that optimistic view. Even at our real-world level, we see that some of the basic questions remain largely unanswered.

The history of science during the three centuries that followed the newtonian synthesis is a dramatic story indeed. There were moments when the program of classical science seemed near completion: a fundamental level, which would be the carrier of deterministic and reversible laws, seemed in sight. However, at each such moment something invariably did not work out as anticipated. The scheme had to be enlarged, and the fundamental level remained elusive.

Today, wherever we look, we find evolution, diversification, and instabilities. We have long known that we are living in a pluralistic world in which we find deterministic as well as stochastic phenomena, reversible as well as irreversible. We observe a great number of deterministic phenomena, such as the frictionless pendulum or the trajectory of the moon around the earth. Moreover, we know that many phenomena—the frictionless pendulum, for one—are also reversible, for future and past play the same role in the equations describing the motion or dynamics involved. But other processes, such as diffusion or chemical reactions, are irreversible. In such processes, there is a privileged direction of time: a system that originally is nonuniform becomes homogeneous in the course of time. In addition, if we want to avoid the paradox of referring the variety of natural phenomena to a program printed at the moment of the Big Bang, we are forced to acknowledge the existence of stochastic processes, those whose dynamics is nondeterministic, probabilistic, even completely random and unpredictable.

What has changed since the beginning of this century is our evaluation of the relative importance of these four types of phenomena: reversible versus irreversible, deterministic versus stochastic.

At the beginning of this century, continuing the tradition of the classical research program, physicists were almost unanimous in agreeing that the fundamental laws of the universe were deterministic and reversible. Processes that did not fit this scheme were taken to be exceptions, merely artifacts due to complexity, which itself had to be accounted for by invoking our ignorance, or our lack of control of the variables involved. Now, at the end of this century, more and more scientists have come to think, as we do, that many fundamental processes shaping nature are irreversible and stochastic; that the deterministic and reversible laws describing the elementary interactions may not be telling the whole story. This leads to a new vision of matter, one no longer passive, as described in the mechanical world view, but associated with spontaneous activity. This change is so deep that we believe we can truly speak of a new dialogue of man with nature.

It is interesting to inquire how such a change could have occurred over a relatively short time span. It is the outcome of unexpected results obtained in quite different areas of investigation in physics and chemistry, such as elementary particles, cosmology, or the study of self-organization in far-from-equilibrium systems.

Who would have believed, fifty years ago, that most—and perhaps all—elementary particles are unstable? Or that we could speak about the evolution of the universe as a whole? Or that far from equilibrium, molecules may "communicate," to use an anthropomorphic term, as witnessed in chemical clocks?

These unexpected discoveries, along with many others, also had a drastic effect on our ideas about the relation between "hard" and "soft" sciences. According to the classical view, there was a sharp distinction between simple systems, such as studied by physics or chemistry, and complex systems, such as studied in biology and the human sciences. Indeed, it would be hard to imagine a greater contrast than the one that exists between the simple models of classical dynamics, or the simple behavior of a gas or a liquid, and the complex processes we discover in the evolution of life or in the history of the human societies.

It is precisely because this gap is narrowing that we now may consider applying new knowledge to situations for which the concepts of classical physics were insufficient or inappropriate, or even essentially meaningless.

In classical physics, the investigator is outside the system that he observes. He is the one who can make independent decisions, while the system itself is subject to deterministic laws. In other terms, there is a "decider" who is "free," and the members of the system, be they individuals or organizations, who are not "free" but must conform to some master plan.

Today we are getting farther and farther away from such a dichotomy. Not only in human sciences, but in physics as well we know that we are both actors and spectators—to use a well-known expression by Niels Bohr. In the place of a construction in which the present implies the future, we have a world in which the future is open, in which time is a construction in which we may all participate.

Complexity in Nature

What is the difference, if any, between the swinging of a pendulum and the beating of the heart, or between a crystal of ice and a snowflake? Is the world of physical and chemical phenomena basically a simple and predictable world where all observed facts can be interpreted adequately by appealing to a few fundamental interactions? Is complexity to be found only in biology? These and related questions will be raised throughout this chapter. An examination of experimental data will lead us to the conclusion that the distinction between physico-chemical and biological phenomena, between "simple" and "complex" behavior, is not as sharp as we might intuitively think. This observation will suggest a *pluralistic view* of the physical world, a view that can accommodate a world in which different kinds of phenomena coexist side by side as the conditions to which a given system is submitted are varied. The view of such an open world is the principal message of this book.

1.1 WHAT IS COMPLEXITY?

Complexity is an idea that is part of our everyday experience. We encounter it in extremely diverse contexts throughout our lives, but most commonly we get the feeling that complexity is somehow related to the various manifestations of life. Thus, to most of us the phenomena dealt with in traditional physics textbooks—such as the free fall of an object under the influence of gravity, or the motion of a pendulum—are fundamentally "simple." They seem simple because apparently only one object and one action (or a very few) are involved. In contrast, we think of our economic system, our language, the mammalian brain, or even the humblest of the bacteria as "complex" systems, perhaps because a great number of interacting elements are involved.

But what about a cubic centimeter of a gas, or a liquid such as water? Here also we are in the presence of systems containing a huge number of interacting elements, the molecules. It is difficult to conceive the magnitude of this number. From elementary physical chemistry we know that for a pressure of 760 millimeters of mercury and a temperature of 0°C, one mole of any gas contains a number of molecules equal to Avogadro's number, 6.023×10^{23}. This is about 10^{19} molecules crowded into a space of one cubic centimeter, 0.06 cubic inch—10 billion billion molecules, moving in all possible directions and continuously colliding with one another.

Is this sheer vastness of number enough to qualify this system as "complex"? Intuition continues to tell us "No," because we do not perceive any coordinated activity or form or dynamics. All those randomly colliding molecules don't seem to be "doing anything," they are not "going anywhere." As a matter of fact we tend to regard such a system as the prototype of disordered, erratic behavior frequently referred to by physicists as *molecular chaos*, in which molecules move incoherently because they cannot recognize each other over distances larger than a few angstroms (Å; $1 \text{ Å} = 10^{-8}$ cm, one hundred millionth of a centimeter). This incoherence of molecular movement, represented in Figure 1(a), results from a basic property of the intermolecular forces of interaction in this and many other systems encountered in nature, namely their *short-range* character.

Suppose now that we submit a cubic centimeter of water to the conditions that prevail during a winter storm. We will obtain an intricately patterned snowflake, perhaps like that in Figure 1(b), with characteristic dendritic form. Now, contemplating the inimitable artwork that nature has performed, we might be ready to speak of complexity. We see that the same system can appear in different aspects, evoking successive impressions of "simplicity" and "complexity."

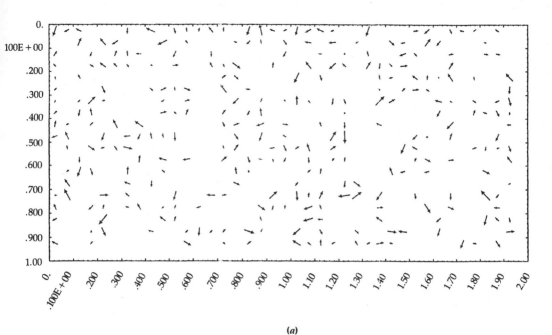

(a)

(b)

Figure 1 (a) Visualization of molecular chaos. The figure summarizes the results of a computer simulation of the equations of motion of 400 elastic hard disks in a rectangular box measuring about 120×60 (molecular diameter)^2. The boundary conditions are temperature on the two horizontal boundaries, fixed at the same value ($T = 1$ in units in which $k_B/m = 1/2$, where k_B is Boltzmann's constant and m is the molecular mass), and periodic conditions on the vertical boundaries. The system was integrated during about 500 collisions per particle, and has attained thermal equilibrium for all practical purposes. The arrows begin at the position of the center of each particle. Their direction and length indicate, respectively, the direction and the magnitude of the velocity of the corresponding particle. (Courtesy of F. Baras). (b) A snow crystal displaying characteristic dendritic form.

This example teaches us an important lesson: It is more natural, or at least less ambiguous, to speak of *complex behavior* rather than complex systems. The study of such behavior will reveal certain common characteristics among different classes of systems and will allow us to arrive at a proper understanding of complexity. That is our goal.

1.2 SELF-ORGANIZATION IN PHYSICO-CHEMICAL SYSTEMS: THE BIRTH OF COMPLEXITY

Science is above all an experimental enterprise, and a significant part of experimentation is observation, so let us look around to get an idea of the generality and importance of complex phenomena.

We have already pointed out that the fascination exerted by biology is responsible for a somewhat diffuse identification of the idea of complexity with the phenomena of life. Surprisingly, this perennial idea will be the first to break down as we pursue an understanding of complexity. Since the 1960s a revolution in both mathematical and physical sciences has imposed a new attitude in the description of nature. Parallel developments in the thermodynamic theory of irreversible phenomena, in the theory of dynamical systems, and in classical mechanics have converged to show in a compelling way that the gap between "simple" and "complex," between "disorder" and "order," is much narrower than previously thought. Simple examples of mechanical systems are now known to present complex behavior. A periodically forced pendulum—such as a playground swing—at the borderline between vibration and rotation gives rise to a rich variety of motions, including the possibility of random turbulentlike excursions from its equilibrium position. Such ordinary systems as a layer of fluid or a mixture of chemical products can generate, under certain conditions, *self-organization phenomena* at a macroscopic scale in the form of spatial patterns or temporal rhythms. In short, the idea of complexity is no longer limited to biology. It is invading the physical sciences and appears to be deeply rooted in the laws of nature.

As a result of these discoveries, interest in macroscopic physics, the physics dealing with phenomena on our scale, is increasing enormously. In the next few sections we describe some particularly representative phenomena to which we will refer continuously throughout the book.

1.3 THERMAL CONVECTION, A PROTOTYPE OF SELF-ORGANIZATION PHENOMENA IN PHYSICS

We shall first consider the bulk motions of fluids (liquids or gases) under the effect of temperature inhomogeneities, motions known as *thermal convection*.

The study of these motions is far from academic. Thermal convection is the basis of several important and spectacular phenomena on our planet. One example is the circulation of atmosphere and oceans, which determines to a large extent short- and medium-term weather changes; another example is continental drift, the motion of continental plates induced by large-scale movements in the mantle. Convection is also the basis of the transfer of heat and matter in the sun, which in turn affects solar activity considerably. In the laboratory, we can study the mechanism of thermal convection with a setup of more modest dimensions than the continents or the solar system. The following simple experiment, first performed in 1900 by the French physicist Bénard, leads to the observation of a number of astonishing properties.

Imagine a layer of fluid, say water, between two horizontal parallel plates whose dimensions are much larger than the width of the layer. Left to itself, the fluid will rapidly tend to a homogeneous state in which, statistically speaking, all its parts will be identical. A minute observer within the fluid, as in Figure 2, will be unable to know on the sole basis of observations of his environment whether he is within the small volume V_A rather than the small volume V_B of the fluid. All volumes that could arbitrarily be defined within the fluid will be indistinguishable, thus knowledge of the state of one of them is also knowledge of the state of any other one, or of them all. In other words, the position our tiny observer occupies makes no difference. And because all positions, all volumes, are identical, and his range of vision never extends to the fluid border, there is no intrinsic way for him to perceive the idea of space.

The homogeneity of this system extends to all its properties, and in particular to its temperature, which will be the same at all parts of the fluid and equal to the temperature of the limiting plates or, alternatively, to the temperature of the external world.

All these properties are characteristic of a system in a particular state, the state for which there is neither a bulk motion nor a temperature difference with the outside world, a state of *equilibrium*. We can express this as follows. We denote the temperatures of plates 1 and 2 by T_1 and T_2, respectively. The temperature

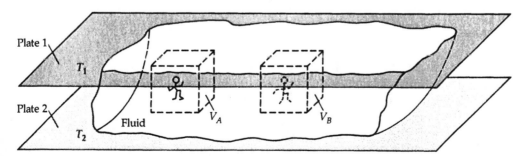

Figure 2 A minute observer contemplating the states of the volume elements V_A, V_B within a thin layer of liquid in equilibrium finds them indistinguishable and concludes that the fluid displays translational invariance along the horizontal direction.

difference, ΔT, is defined $\Delta T = T_2 - T_1$, and since at equilibrium there is no temperature difference, we have:

$$\Delta T_e = T_2 - T_1 = 0 \qquad (1.1)$$

Imagine now that someone briefly touches his finger to one plate. The temperature in this part of the plate will momentarily be modified, for instance, from 20°C to the human body's temperature of 36.5°C. An incident like this, one that takes place by chance in a system and locally (and generally weakly) modifies some of its properties, is called a *perturbation*. For our system at equilibrium this temperature perturbation will have no influence, since once the finger is removed the temperature will rapidly become uniform again and equal to its initial value of 20°C. In other words the perturbation dies out; the system keeps no track of it. When a system is in a state such that the perturbations acting on it die out in time more or less quickly, we say that the state is *asymptotically stable*.

From the standpoint of our minute observer, the stability of the state of equilibrium makes all instants identical. It is therefore also impossible for him to develop an intrinsic conception of time, for the perception of time involves an awareness of change, and there is no change. One can hardly speak of "behavior" for a system in such a *simple* situation; it just sits there "doing nothing."

We can induce behavior by heating the fluid layer, say from below. In doing so we communicate energy to the system in the form of heat. As the temperature T_2 of the lower plate becomes higher than T_1, the equilibrium condition in Eq. (1.1) is violated ($\Delta T > 0$). In other words, by applying an *external constraint* we do not permit the system to remain at equilibrium. Note that in the present example an external constraint implies energy flux and vice versa.

Suppose that the constraint is weak (ΔT small). The system will adopt a simple and unique state in which heat is transported from the lower to the upper plate, where it is evacuated to the external world. The only difference, compared with the state of equilibrium, is that the temperature, and through it the density and pressure in the fluid, will no longer be uniform. These properties will differ in practically linear fashion, from warm regions below to colder regions above. This phenomenon is known as *thermal conduction*. In this new state that the system has reached in response to the constraint, stability again prevails and the behavior is as "simple" as at equilibrium.

If we remove the system farther and farther from equilibrium by increasing ΔT, suddenly, at a value of ΔT that we will call *critical*, ΔT_c, matter—the fluid—begins to perform a bulk movement. Moreover, this movement is far from random: the fluid is structured in a series of small convection "cells" represented in Figure 3, known as Bénard cells. The fluid is now in the regime of thermal convection.

Figure 4 shows a qualitative explanation of the phenomenon. Owing to thermal expansion the confined fluid layer becomes stratified, with the part close to the lower plate characterized by a lower density than the upper part. This gives rise

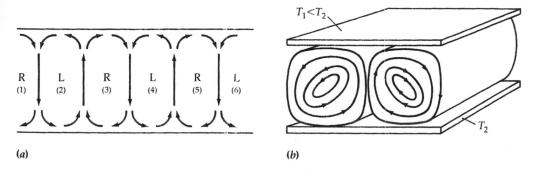

(a)

(b)

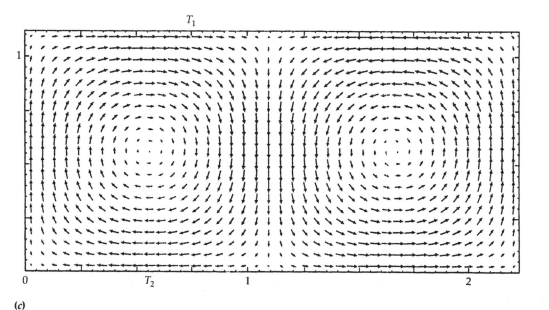

(c)

Figure 3 (a)(b) Two views of convection (Bénard) cells. Notice the direction of rotation in any two adjacent cells, further represented in Figure 3(c). (c) Complexity and long-range order out of molecular chaos in a system under nonequilibrium constraint. The equations of motion of 5000 elastic hard discs in a rectangular box measuring about 112×56 (molecular diameter)^2 were solved by computer with the following boundary conditions: fixed temperatures at the lower and upper horizontal boundaries—respectively, $T_2 = 1.61$, $T_1 = 0.51$ (in reduced units in which the sound velocity in the medium is equal to one)—and perfectly reflecting vertical walls. The system was first integrated for a time interval corresponding to the occurrence of over 30×10^6 collisions. Subsequently the instantaneous molecular velocities were recorded for 20×10^6 more collisions and their averages over this time interval as well as over a small spatial cell containing five or six molecules were plotted. The significance of the arrows is as in Figure 1(a). Notice that in the equilibrium situation of Figure 1(a) a similar averaging would give rise to a zero velocity field everywhere. (From M. Mareschal, M. Malek Mansour, A. Puhl, and E. Kestemont, *Phys. Rev. Letters* **61**, 2550 [1988].)

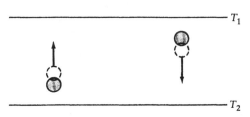

Figure 4 Qualitative explanation of the origin of thermal convection. If a parcel of warm fluid near the bottom of the layer is displaced upward slightly, it enters a region of greater average density and therefore experiences an upward buoyancy force. Similarly, a parcel of cool fluid near the top of the layer, if displaced downward slightly, becomes heavier than its surroundings and tends to sink.

to a continuous change in density, a gradient from low to high upward through the fluid that opposes the force of gravity. This configuration is potentially unstable. Consider a small volume of the fluid near the lower plate. Imagine that this volume element is weakly displaced upward by a perturbation. Being now in a colder—and hence denser—region, it will experience an upward Archimedes force that will tend to amplify the ascending movement further. If, on the other hand, a small droplet initially close to the upper plate is displaced downward, it will penetrate an environment of lower density, and the Archimedes force will tend to amplify the initial descent further. We see therefore that, in principle, the fluid can generate ascending and descending currents like those observed in the experiment. The reason these currents do not appear as soon as ΔT is not strictly zero is that the destabilizing effects are counteracted by the stabilizing effects of the viscosity of the fluid, which generates an internal friction opposing movement, as well as by thermal conduction, which tends to smear out the temperature difference between the displaced droplet and its environment. This explains the existence of a critical threshold, ΔT_c, observed in the experiment. Figures 3(b) and 3(c) show the complexity of the movements: the fluid goes upward, moves along plate 1, then goes downward, moves along plate 2, goes upward again, and so on. The direction of rotation alternates from cell to cell along the horizontal axis, being successively right-handed (D) and left-handed (L).

The complexity of this convective movement might seem modest compared to the activity of even the humblest bacterium, but that is a matter of scale. Let us turn once again to our minute observer: His universe has been totally transformed. For instance, he can now decide where he is and where he is not by observing the sense of rotation of the cell he occupies. Moreover, by moving along the horizontal from cell to cell and counting the number of cells he goes through, he can acquire a quite efficient notion of distance in space. The emergence of the concept of space in a system in which space could not previously be perceived in an intrinsic manner is called *symmetry breaking*. In a way, symmetry breaking brings us from a static, geometrical view of space to an "Aristotelian" view in which space is shaped or defined by the functions going on in the system.

Perhaps the most remarkable feature to be stressed in the sudden transition from simple to complex behavior is the *order* and *coherence* of this system. When ΔT is below the critical value ΔT_c, the homogeneity of the fluid in the horizontal direction makes its different parts, which can only be defined arbitrarily, independent of each other. It is totally without importance if the positions of volumes V_A and V_B in Figure 2 are interchanged. If there were a volume V_C between V_A and V_B, no change in its observable properties could be detected simply because volume A was now on its right rather than on its left. But beyond the threshold ΔT_c, everything happens as if each volume element was watching the behavior of its neighbors and was taking it into account so as to play its own role adequately and to participate in the overall pattern. This suggests the existence of *correlations*, that is, statistically reproducible relations between distant parts of the system. This key conclusion is reached here just by observing the phenomenon of transition across ΔT_c. The whole subject will be taken up again and analyzed in detail in Chapters 2 and 4. However, even at this early stage of our analysis it is important to note the long-range character of these correlations as compared to the short range of the intermolecular forces. The characteristic space dimension of a Bénard cell in usual laboratory conditions is in the millimeter range (10^{-1} centimeter), whereas the characteristic space scale of the intermolecular forces is in the angstrom range (10^{-8} centimeter). Intermolecular forces operate up to a distance equal to about one molecule; a single Bénard cell comprises something like 10^{20} molecules. That this huge number of particles can behave in a coherent fashion, as in the case of convective flow, despite the random thermal motion of each of them is one of the principal properties characterizing the emergence of complex behavior. Figure 3(c) provides us with a visualization of the emergence of this "organized" complexity out of the interplay between the disordered thermal motion of the individual molecules and the nonequilibrium constraint. The comparison with the molecular chaos visualized in Figure 1(a) is striking.

In the preceding discussion we have been using a vocabulary that includes concepts such as coherence, complexity, and order. These have long been an integral part of biology, but until recently they were outside the mainstream of physics. The possibility of using these fundamental concepts to describe the behavior of quite ordinary physical systems as well as living things is a major development that science could not have forecast even just a few years ago.

Bénard cells hold more surprises for us. The experiment that produces Bénard cells is perfectly reproducible: as long as the experimental conditions—fluid, plates, plate spacing—are the same, the convection pattern will appear at the same threshold value ΔT_c. As shown in Figure 3, the fluid becomes structured in cells that are alternately right-handed and left-handed, and once a direction of rotation is established in a particular cell, it will not change. However, no matter

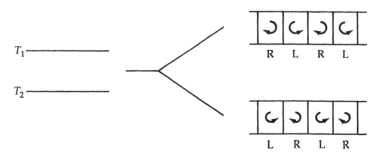

Figure 5 Multiplicity of solutions beyond the threshold of the thermal convection instability. Depending on the conditions, a given region of space may be part of a left-handed or a right-handed cell.

how sophisticated the control of the experimental setup might be, two qualitatively different situations can be realized just after the critical threshold ΔT_c.* In Figure 3(a), the odd-numbered cells are right-handed and the intervening, even-numbered cells are left-handed. But the opposite pattern of alternate left and right rotation also occurs. The two patterns are compared in Figure 5. In short, as soon as ΔT slightly exceeds ΔT_c, we know that the cells will appear: this phenomenon is therefore subject to a strict determinism. In contrast, the *direction* of rotation of the cells is unpredictable and uncontrollable. Only chance, in the form of the particular perturbation that may have prevailed at the moment of the experiment, will decide whether a given cell is right- or left-handed. We thus arrive at a remarkable cooperation between chance and determinism, one that is reminiscent of the duality of mutation (chance) and natural selection (determinism) familiar in biology since Darwin's era, and which was previously limited in the physical sciences to the quantum description of phenomena going on at a microscopic scale.

What we have seen above is this: Far from equilibrium, that is, when a constraint is sufficiently strong, the system can adjust to its environment in several different ways. Stated more formally, *several solutions* are possible for the *same parameter values*. Chance alone will decide which of these solutions will be realized. The fact that only one among many possibilities occurred gives the system a *historical dimension*, some sort of "memory" of a past event that took place at a critical moment and which will affect its further evolution.

What happens when the thermal constraint increases well beyond the first threshold at which structuring occurs? For some range of values the Bénard cells will be maintained globally, but some of their specific characteristics will be

---

As a matter of fact one should say that there is an infinity of possibilities, since for a large system the whole structure can be moved rigidly along the horizontal direction (breaking of a continuous symmetry group). However, the most characteristic manifestation of this multiplicity remains the direction of rotation of the cell, and for this reason we use this criterion throughout the present section.

modified. Then suddenly, beyond another critical value $\Delta T_c{}'$, a new forceful manifestation of randomness appears: the structure becomes fuzzy and a regime characterized by an erratic dependence of the variables on time emerges. This is the precursor of what, for about a century, engineers and fluid dynamicists have been calling *turbulence*. It now appears that turbulence is one aspect of a general trend of several classes of systems to evolve in a *chaotic* fashion under certain conditions. We discuss this point further in Chapter 3.

To summarize, we have seen that nonequilibrium has enabled the system to avoid the thermal disorder depicted in Figure 1(a) and to transform part of the energy communicated from the environment into an ordered behavior of a new type, the *dissipative structure*: a regime characterized by symmetry breaking, multiple choices, and correlations of a macroscopic range. We can therefore say that we have witnessed the birth of complexity. True, the type of complexity achieved is rather modest; nevertheless, it presents characteristics that usually have been ascribed exclusively to biological systems. More important, far from challenging the laws of physics, complexity appears as an inevitable consequence of these laws when suitable conditions are fulfilled.

1.4 SELF-ORGANIZATION PHENOMENA IN CHEMISTRY

We have followed tradition in regarding Bénard cells as part of physics, principally because the chemical nature of the substances constituting the confined fluid layer remains unchanged during the phenomenon. We shall now consider chemical reactions—processes that involve modifications of the identity of the molecules of the reacting substances. Such phenomena are of quite general concern. Much of the chemical industry is based on heterogeneous catalysis, in which some of the steps necessary for the synthesis of a product are accelerated by the presence of a surface reacting with the bulk medium. For instance, the oxidation of ammonia is usually carried out in the presence of a platinum catalyst; similarly, in the decomposition of nitrous oxide, N_2O, a catalytic surface of copper oxide is used. Combustion, the burning of hydrocarbons, the process by which heat engines function, is another class of important chemical transformations. In addition, several manifestations of biological activity are amenable to chemical reactions involving special catalysts, the enzymes.

In a typical chemical reaction a molecule of species A (say the hydroxyl radical OH) can combine with a molecule of species B (say molecular hydrogen H_2) to produce one molecule of species C and one molecule of species D (respectively H_2O and atomic hydrogen H in our example). This process is symbolized

$$A + B \xrightarrow{\ k\ } C + D \tag{1.2a}$$

in which k is the rate constant, generally a function of temperature and pressure. On the lefthand side, the *reactants* A and B combine and disappear in the course of time, whereas on the righthand side the *products* C and D are formed and appear as the reaction advances. In an isolated system, even after a very long time reactants A and B never disappear completely. More precisely, after a sufficient reaction time the amounts of the coexisting constituents A, B, C, and D attain a fixed value of the ratio $c_C c_D / c_A c_B$, where c denotes the concentration of each constituent. When this fixed value is attained we say that the system is at *chemical equilibrium*, and the value of the ratio is the *equilibrium constant*. This is the analog of the homogeneous state of rest in the Bénard problem. It is of course understood that the values of the physical parameters such as pressure and temperature remain constant.

How can one reconcile this result with Eq. (1.2a), according to which A and B are bound to disappear? Experiment shows the existence of the reverse transformation of Eq. (1.2a),

$$C + D \xrightarrow{k'} A + B \tag{1.2b}$$

At equilibrium both reactions occur with exactly the same velocity. This fundamental property of nature, known as *detailed balance*, is responsible for most of the properties characterizing this state of matter. We represent a reversible reaction as

$$A + B \underset{k'}{\overset{k}{\rightleftarrows}} C + D \tag{1.2c}$$

In the Bénard problem we were able to move the system away from equilibrium by communicating an energy flux to it. The most obvious chemical analog of this procedure is to submit the system to a flow of mass from (or toward) the surroundings, thus producing what is known as an *open system*. For instance, we can eliminate C or D from the reaction vessel when their concentration becomes larger than a prescribed value; or we can feed the system by a flow of a mixture rich in A and B that goes through the reaction vessel and is eventually evacuated or recycled.

Figure 6 depicts the setup for a laboratory-scale open system. By combining appropriate rates of inflow and outflow of matter, we can create conditions for the system to attain a state in which the concentrations of A, B, C, and D remain constant in time while at the same time their ratio is no longer given by the equilibrium constant. Mathematically, this will be reflected as a vanishing rate of change of the concentrations in time. To express this properly we introduce the time derivatives of the concentration and write

$$\frac{dc_A}{dt} = \frac{dc_B}{dt} = \cdots = 0 \tag{1.3}$$

We call the state described above a *stationary nonequilibrium state*.

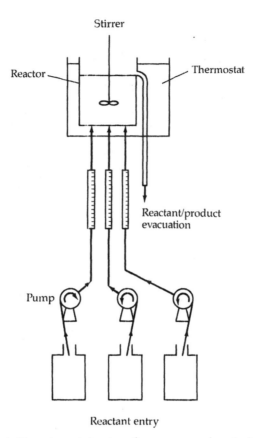

Figure 6 Experimental setup for an open chemical system.

Can we expect from an open system around such a nonequilibrium state a behavior similar to the Bénard problem? For one thing, in such a system detailed
balance no longer holds. Suppose now that one particular process, say the forward
step of a reaction, momentarily becomes enhanced compared to the backward
step. At equilibrium, detailed balance would tend to reestablish the initial state of
affairs. Away from equilibrium, however, this is no longer so. Moreover, if some
parts of the chemical mechanism could enable the system to capture and further
amplify the enhancement, we would have a potentially unstable situation similar
to the one depicted for the Bénard problem in Figure 4. Such mechanisms are
known to exist in chemistry, and their most striking manifestation is *autocatalysis*.
For instance, the presence of a product may enhance the rate of its own production. As a matter of fact this seemingly exotic phenomenon happens routinely in
any combustion process, thanks to the presence of free radicals, those extremely
reactive substances containing one unpaired electron, which by reacting with
other molecules give rise to further amounts of free radicals and thus to a self-
accelerating process. In addition, *self-reproduction*, one of the most characteristic

properties of life, is basically the result of an autocatalytic cycle in which the genetic material is replicated by the intervention of specific proteins, themselves synthesized through the instructions contained in the genetic material.

For a long time chemists thought that a homogeneous, time-independent state similar to equilibrium should eventually emerge from any chemical transformation. However, a chemical reaction known as the Belousov-Zhabotinski (BZ) reaction under certain conditions of nonequilibrium presents a whole spectrum of fascinating and unexpected behaviors.

There is nothing particularly unusual about the reactants of the BZ reaction. A typical preparation consists of cerium sulfate $Ce_2(SO_4)_3$, malonic acid $CH_2(COOH)_2$, and potassium bromate $KBrO_3$, all dissolved in sulfuric acid. The evolution of the system can be followed visually since an excess of Ce^{4+} ions gives a pale yellow color, whereas an excess of Ce^{3+} ions leaves the solution colorless. More sophisticated observations can be obtained by placing electrodes in the solution or by making spectroscopic measurements of the optical absorption caused by a particular substance.

There are several types of behavior exhibited by this system under different experimental conditions, all at ordinary temperature; they include clock behavior, turbulence, spatial patterning, and hysteresis.

BZ reaction in a well-stirred system: Chemical clock and chaos

Suppose that the reaction is first carried out with the setup shown in Figure 6. Thanks to the efficient transfer of matter ensured by stirring, the system—that is, the mixed solution—remains practically homogeneous in space at each moment. This experimental setup allows easy control over the distance of the system from equilibrium; simply changing the rates at which the chemicals are pumped into or out of the system varies the *residence times* of these substances within the reaction vessel. Very long reactant residence times essentially result in a closed system, and under these conditions we expect the system to reach an equilibriumlike behavior characterized by detailed balance. Conversely, by decreasing reactant residence times we do not permit a full equilibration of forward and backward steps, and under those conditions we expect the system to manifest nonequilibrium behavior. This is precisely what the experiment shows. For very large residence times a homogeneous steady state is reached in which the concentrations of the chemicals remain time-independent. This is the typical state to which the laboratory chemist is accustomed, and it shares all the qualitative properties of chemical equilibrium. It is the analog of the state of thermal conduction achieved in the Bénard system when a weak temperature difference is applied across the plates.

If we now reduce the residence time, we encounter an altogether different pattern of behavior. Suddenly a pale yellow color indicating an excess of Ce^{4+} ions invades the system. A few minutes (or according to the circumstances even

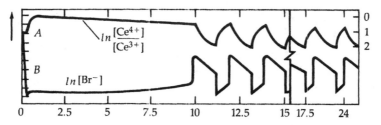

Figure 7 Potentiometric traces of ln Br$^-$ and ln (Ce^{4+}/Ce^{3+}) versus time during the Belousov-Zhabotinski reaction. Initial concentrations: CH$_2$(COOH)$_2$ = 0.032 M; KBrO$_3$ = 0.063 M; Ce(NH$_4$)$_2$(NO$_3$)$_5$ = 0.01 M; H$_2$SO$_4$ = 0.8 M; KBr = 1.5 × 10^{-5} M.

a fraction of a minute) later the solution becomes colorless, indicating an excess of Ce^{3+} ions. The process will go on, yellow, colorless, yellow, colorless . . . , in a rhythm that has a perfectly regular period and amplitude which depend only on the experimental parameters and are thus intrinsic to the system. This oscillation, which measures time through an internally generated dynamics, constitutes a *chemical clock*. Figure 7 shows a record of typical clock behavior.

But why are these oscillations noteworthy? What is significant in the rhythmic behavior of a chemical clock? Isn't it the analogue of well-known physical behavior, the swinging of a pendulum?

Figure 8 illustrates the differences between the two kinds of oscillations. The upper left diagram shows a frictionless pendulum swinging with a maximum angle

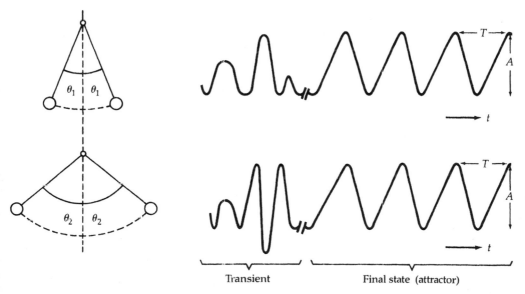

Figure 8 Comparison of sustained oscillations. Left, conservative system, the pendulum; right, dissipative system, the BZ reagent.

to the vertical of opening θ_1 [the amplitude of the periodic function $\theta(t)$, describing the instantaneous value of the angle] and a period T_1. The upper right diagram shows the variation in time of the concentration of a chemical in the BZ reagent, irregular at first, and then characterized by regular oscillations of amplitude A and a period T when the clocklike state has been attained.

We now momentarily disturb the situation by displacing the pendulum at a higher angle θ_2 from the vertical $(\theta_2 > \theta_1)$, and by applying a slight concentration or temperature pulse in the BZ reagent—for instance, by injecting a few additional millimoles of potassium bromate or by touching the vessel for a few seconds. The lower half of Figure 8 describes the response of the systems to these brief disturbances. The pendulum will continue to oscillate, but now its amplitude will be equal to θ_2 rather than θ_1 and its period will be slightly larger. In other words, this system will keep the memory of the disturbance forever. In contrast, after a transient the BZ reagent will resume an oscillatory mode of exactly the same amplitude and period as before. This is the property of *asymptotic stability* to which we referred earlier. It is directly related to *irreversibility*, a property of most of the phenomena observed in nature and the one that is essentially responsible for the reproducibility of events. Systems such as the pendulum lack this property because their dynamics is invariant with time reversal. They are therefore at the complete mercy of the perturbations that may act on them, and as these perturbations are essentially unpredictable, such systems are bound to show erratic behavior sooner or later.

In Chapter 3 we will make a more systematic study of the differences between *dissipative systems*, systems that display irreversibility, like Bénard flow and the BZ reaction, and *conservative systems* like the pendulum, in which dynamics is invariant with time reversal. For now, one more example will underline the universal importance of dissipative systems. An oscillator with which each of us is intimately involved throughout our lives is the human heart. It beats more or less regularly at about 70 or 80 beats per minute for the average individual, but it can become deregulated. A particularly dramatic form of deregulation is fibrilation, the sudden inability to perform the complete set of muscular actions that constitute an entire heart beat. (Whatever the ultimate cause, the death of an individual always involves a state of fibrilation of the heart.) Suppose now that because of some perturbation, the pattern of oscillation—the normal heart beat rhythm—is upset. Since the human system is subject to a great many perturbations every day, if the heart functioned as a pendulum does fibrilation could well have occurred in the embryo, before birth. But the heart is not like a pendulum, it does not "remember" the effect of a perturbation by permanently changing its pattern of oscillation; if no permanent physical damage has occurred and the cause of the perturbation is removed, the heart resumes its normal rhythm. This is true of any other reproducible phenomenon observed in nature, from circadian rhythms and the cell division cycle to the change of luminosity of variable stars, the cepheids. The fact that these and all other reproducible rhythmic phenomena observed in nature belong to the same realm as the oscillations in the BZ reagent shows the tremendous importance of irreversibility and dissipative systems.

Let us return to our chemical clock. In the vocabulary introduced in Section 1.3, we can say that in the regime of uniform steady state (which is also asymptotically stable) the system ignores time. But once in the periodic regime, it suddenly "discovers" time in the phase of the periodic motion and in the fact that the maxima of the different concentrations follow each other in a prescribed order. We refer to this as the *breaking of temporal symmetry*.

From an even more fundamental viewpoint, the maintenance of sustained oscillatory behavior encompassing the entire system implies that its different parts act in a concerted fashion by maintaining sharp phase relationships between themselves; otherwise, destructive interference would wipe out the oscillatory behavior. In other words, we again expect to observe, just as in the Bénard problem, the emergence of *long-range correlations* induced by the nonequilibrium constraints. In Chapters 2 and 4 the onset and the characteristics of such correlations will be analyzed in considerable detail.

But we have not yet exhausted the list of surprises hidden in the BZ reaction. Detailed experimental analysis shows that for chemical residence times that are intermediate between the times associated with two kinds of oscillatory regimes, complex nonperiodic behavior is observed. The appearance of this *chemical turbulence* illustrates once again the tendency of many natural systems to evolve in a chaotic way under certain conditions. It also brings out the following very important property of chemical systems. In fluid mechanics and in most other physical examples, complex behavior is invariably associated with spatial inhomogeneities. But in chemistry even a spatially homogeneous system can show complex behavior in time. The reason for this, mentioned previously, is that chemical systems are endowed with mechanisms, such as autocatalysis, that are due to the peculiar molecular structure and reactivity of certain constituents and which enable them to evolve to new states by amplifying (or repressing) the effect of slight perturbations.

BZ reaction in a nonuniform system: Spatial patterns

Suppose now that the BZ reaction is carried out without stirring, thus allowing for the possible development of spatial inhomogeneities. We will then see regular patterns in space and time in the form of propagating wave fronts. Still photographs taken at various instants, as in Figure 9 give only a faint idea of the beauty and activity involved in the evolution of this reaction. The waves seen in the figure are created in a thin layer of reagent. They appear primarily in two different forms: circular fronts (a) displaying a roughly cylindrical symmetry around an axis perpendicular to the layer, usually referred to as *target patterns*; and spiral fronts (b) rotating in space clockwise or counterclockwise. It is also possible to obtain, although under rather exceptional conditions, the multiarmed spirals shown in

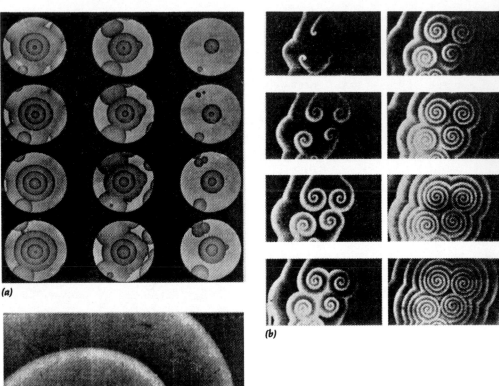

Figure 9 Wave propagation in a two-dimensional layer of BZ reagent. (a) Target patterns. (b) Spiral waves. (c) Multiarmed spirals.

Figure 9(c). In each case the wave fronts propagate over macroscopic distances of space, without distortion and at a prescribed speed; they both represent and transmit the "message" released by the chemistry at the center from which the whole pattern emanates. Once again, we witness the birth of complexity.

As in the Bénard problem, we can associate the formation of wave fronts with a space symmetry breaking. The symmetry broken by the target patterns of Figure 9(a) is quite similar to that broken in the Bénard system: essentially, the system loses invariance to translations along a particular direction of space. The symmetry breaking involved in the formation of spirals, Figures 9(b) and (c), is quite different, as it is associated with *chirality*, or rotation. This type of asymmetry of matter has always exerted a particular fascination. Louis Pasteur, the founder of modern biochemistry, repeatedly expressed his bewilderment over the optical asymmetry of biomolecules, which is manifested by the rotation of the plane of polarization of light in a preferred direction. He regarded this property as one of the basic aspects of life. Moreover, the observation of the morphological asymmetry of adult organisms has introduced into human thinking the notions of "right" and "left" which have influenced philosophers and writers ever since Plato. It is amazing to see these deep notions emerging quite naturally through the intrinsic dynamics of a modest, ordinary-looking physico-chemical system.

The behavior we have described is not an exclusive feature of the BZ system. It is shared by a host of other reactions that occur in a homogeneous phase and involve equally simple chemicals; a partial list is given in Table 1. Characteristically, in all these situations the domain of parameter values for which oscillations are observed is quite close to the domain in which another interesting phenomenon occurs, namely *bistability*. Specifically, two (or sometimes several)

Table 1 Chemical oscillators

Main species	Additional species
In a continuous stirred tank reactor (CSTR)	
I^-	IO_3^-, MnO_4^-, or $Cr_2O_7^-$
I^-	Malonic acid
IO_3^-	H_3AsO_3
IO_3^-	$Fe(CN)_6^{4-}$, SO_3^{2-}, ascorbic acid, or $CH_2O \cdot SO_2$
I_2	$Fe(CN)_6^{4-}$, SO_3^{2-}
IO_3^-	I^-, malonic acid
IO_3^-	I^-, H_3AsO_3
I^-	BrO_3^-
BrO_3^-	SO_3^{2-}, $Fe(CN)_6^{4-}$, H_3AsO_3, or Sn^{2+}
I^-	I_2, $S_2O_3^{2-}$

Main species	Catalysts
In surface catalysis	
$CO + O_2$	Pt, Pd, CuO, Ir
$H_2 + O_2$	Pt, Pd, Ni
$NH_3 + O_2$	Pt
$C_2H_4 + O_2$	Pt
$C_3H_6 + O_2$	Pt
$C_6H_{12} + O_2$	NaY (zeolite)
N_2O (decomposition)	CuO

simultaneously stable stationary states coexist under exactly the same boundary conditions. The particular state chosen by the system will depend on the experimental conditions. This is illustrated in Figure 10, where the value of a variable such as the concentration of a chemical is plotted versus a characteristic control parameter λ, for example, the residence time. Suppose that an experiment is performed for a value $\lambda = \lambda'$ of this parameter, for which only one stable state, branch (a) is available. By gradually increasing λ we enter in region of multiple states ($\lambda_1 < \lambda < \lambda_2$). However, the system remains on branch (a) until λ exceeds the value λ_2. At this moment it jumps to branch (b) and remains there. If a variation of λ in the opposite direction is now imposed, starting say from value λ'' and going to value λ', the system will remain on branch (b) until the parameter reaches value λ_1, whereupon it will jump to branch (a). In other words, the system describes different patterns according to its past history. We call this phenomenon *hysteresis*.

In all cases known so far, it turns out that the same chemical mechanism can account for bistability, oscillations, and waves. As already mentioned, at least one autocatalytic step is usually involved. In the BZ system this step ensures the production of two molecules of an intermediate substance, the bromous acid $HBrO_2$,

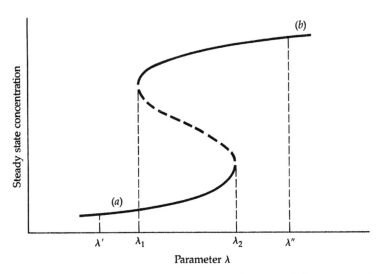

Figure 10 Illustration of the phenomena of *bistability* and *hysteresis*. The system is started on branch (a) with an initial value of parameter λ equal to λ'. Next, λ is increased. The system first remains on (a), but beyond the limit point value $\lambda = \lambda_2$, it jumps to branch (b). When λ is decreased below the value λ'', the system remains on branch (b) until λ becomes λ_1, whereupon it jumps to (a). In other words, the specific path of states followed depends on the system's past history. This is known as hysteresis. Notice that in the range $\lambda_1 < \lambda < \lambda_2$ the system has bistability—that is, the possibility to evolve, for given parameter values, to more than one stable state, depending on initial preparation.

out of one molecule of the same substance according to

$$HBrO_2 + BrO_3^- + 3H^+ + 2Ce^{3+} \rightarrow 2HBrO_2 + 2Ce^{4+} + H_2O \quad (1.4)$$

Direct autocatalysis is not the only mechanism capable of generating complex behavior in chemistry. For instance, in the phenomenon of heterogeneous catalysis some of the chemical steps usually release energy, which heats the medium; for this reason they are called exothermic reactions. Now the rate constant of a chemical reaction is temperature-dependent. It is in fact an increasing function of temperature, described to a good approximation by the Arrhenius law:

$$k(T) = k_0 e^{-E_0/k_B T} \quad (1.5)$$

in which E_0 is referred to as the activation energy and k_B is a universal constant known as Boltzmann's constant. This dependence can be understood qualitatively as follows, and one possibility is shown in Figure 11. A reactive transformation involves the breaking of a chemical bond, that is, it has to overcome an "energy barrier" corresponding to the energy of the bond. The required amount of reaction energy is provided by the kinetic energy of translational motion of the colliding molecules. If the medium is heated, the mean kinetic energy of the molecules will increase, and thus a greater number of pairs of colliding molecules will have a sufficient amount of kinetic energy to overcome the barrier. In other words the reaction will be accelerated, in agreement with Eq. (1.5).

Suppose now that we are dealing with an exothermic reaction. If such a process is accelerated, more heat will be released, the temperature of the medium will further increase, and the reaction will be further accelerated. This provides a potentially destabilizing element capable of inducing transitions to new types of

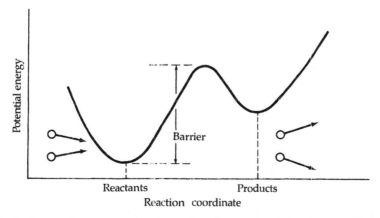

Figure 11 A chemical reaction represented in terms of motion in a double-well potential. The two minima correspond to two stable configurations of the species involved in the process, which are identified as *reactants* and *products*.

behavior. And, indeed, in many catalytic reactions of industrial interest patterns of behavior quite similar to those described for the BZ system are observed. These have considerable effect on the local state of the catalyst and therefore play a very important role in the course of the reaction and in the yield and efficiency of the chemical plant.

1.5 PHYSICO-CHEMICAL COMPLEXITY AND ALGORITHMIC COMPLEXITY

We have now encountered three basic modes of self-organization of matter giving rise to complex behavior: bistability and hysteresis, oscillations (both periodic and nonperiodic), and space patterns. We have seen these phenomena emerge in two cases: bulk flow under a thermal constraint and an open system undergoing catalytic chemical reactions. Let us see whether, with the help of these examples, we can produce a clear-cut definition of complexity. This should be of great help for our subsequent analysis of the mechanisms responsible for the onset of complex behavior.

One way to arrive at a definition capturing the essence of complexity is to make an abstraction of the details of the specific experimental factors considered in Sections 1.3 and 1.4, and try to communicate to another observer a set of instructions that would enable him to reproduce the observed behavior without performing the experiment again. Suppose that the cost of communicating with such an observer was very high. This would force upon us the additional constraint of making the instructions as short as possible without compromising the feasibility of the original goal.

How would we proceed under those circumstances in the case of the Bénard experiment? To describe the equilibrium situation depicted in Figure 2 the following minimal instructions are both necessary and sufficient:*

- $\Delta T = T_2 - T_1 = 0$
- $v = 0$ at any point on a (horizontal) plane parallel to the plates,
  v being the convection velocity

Let us now try to communicate the features at the pattern arising beyond the Bénard instability. Referring to Figures 3 and 5, we realize that the message should now look like this:

- $\Delta T = T_2 - T_1 > \Delta T_c$

---

* *To simplify we focus on data concerning the boundary conditions and the velocity v, leaving aside temperature and pressure.*

- $v = v_0$ at a particular point P chosen as reference point
- This particular point belongs to a cell rotating in a well-defined direction (L or D)
- v at any other point can be deduced from v_0 on the basis of the periodic sequence LDLD . . . , and the value of the wavelength of the pattern

We see that as we move across the Bénard threshold more instructions become necessary in order to specify the state generated by the system. This immediately invokes the definition of *algorithmic complexity*, independently proposed about 1965 by the Soviet mathematician A. Kolmogorov and the American mathematician G. Chaitin. The algorithmic complexity of a sequence of data is defined as the minimum length of a computational algorithm (measured, for instance, in number of bits if the algorithm is to be communicated to a digital computer) that would produce this sequence as output.

Is the algorithmic complexity just defined tantamount to the complexity observed in physical sciences and biology? Consider a sequence N data long expressed in binary form and displaying an overall regularity (e.g., 100 100 100 . . .). Clearly, the message contained in it can be considerably compressed. For instance, it could be transmitted to a computer by the very simple algorithm "Print *100* ten [or 100, or one million] times." The number of binary digits in such an algorithm is a small fraction of the number in the initial series, and as the series grows larger in size the algorithm size increases at a much slower rate. According to the definition, this therefore implies limited algorithmic complexity. Consider now the set of all random sequences N digits long. The number of such sequences is 2^N. Suppose that as N increases some sequences can be compressed as in the previous example, that is, can be expressed in the more compact form of sequences 1, 2, . . . up to K digits long, such that K/N goes to zero as N goes to infinity. The total number of such "orderly" sequences is

$$2 + 2^2 + \cdots + 2^K = 2^{K+1} - 2$$

Compared to 2^N this number is very small when N becomes large. In other words, most of the random sequences will be incompressible. According to the definition this implies a maximum algorithmic complexity that is essentially equal to the sequence length, N. This kind of complexity can be generated in a typical coin-tossing experiment, provided that sufficient care is taken to eliminate all possible correlations between successive steps, thus ensuring a "fair" game.

Now this is far from corresponding to the intuitive view suggested by the experimental results of Sections 1.3 and 1.4, in which complexity is to be associated with long-range coherence and concerted behavior. True, full orderliness in the form of a complete lack of variability is an extreme case of coherence in which the object is like a fossil, and its behavior can hardly be characterized as complex. On the other hand, the strong variability represented by random noise and the concomitant lack of correlations is another, equally nonrepresentative form of organization. In reality physico-chemical complexity must somehow be sandwiched between these two extremes, and thus should not be fully identified with algorithmic complexity.

As we have seen, under certain conditions physico-chemical systems can generate randomness in the form of chaotic dynamics. We shall later show, in Sections 3.10, 4.7, and 4.8, that there are some basic differences between this form of randomness and the randomness generated by noise. Nevertheless, chaotic dynamics will suggest some deep connections between physical and algorithmic complexity; these cannot be fully appreciated at the present stage of our discussion and are therefore postponed until Chapter 4.

1.6 SOME FURTHER EXAMPLES OF COMPLEX BEHAVIOR ON OUR SCALE

As it turns out, the three modes of organization discussed in the preceding three sections are encountered in a host of other macroscopic problems arising in quite different contexts. Without trying to be exhaustive, here are a few characteristic examples; in each case we emphasize the basic causes provoking the transitions.

Surface-tension-induced phenomena: Materials science

Exchanges of energy and matter under nonequilibrium conditions between identical phases of matter (solids or liquids), between phases of different types, or between biological membranes and liquids or gases take place in many circumstances. These processes are characterized by the appearance of *interfaces*, surfaces of separation between the phases that are present. Just as pressure is the coefficient of proportionality between the work done in a change of volume of a body and the volume change itself, so *surface tension* is a coefficient of proportionality between the work necessary for changing the area of a surface element and the area change of this element. As a rule, surface tension depends on curvature, temperature, and composition. It therefore constitutes a coupling mechanism between these quantities and the mechanical properties of the material.

One manifestation of this coupling is a spontaneous surface deformation giving rise to regular spatial patterns on the surface, to bulk flow in the adjacent fluids previously at rest, and to the formation of drops. These phenomena are important in such fields as pharmacology, the food industry, and tertiary oil recovery.

Another striking manifestation arises in the growth of a pure solid or an alloy from the melt, when the progress of the interface is induced by an externally driven freezing front moving toward the melt. In addition to its industrial importance, this process is at work when snow forms from water droplets. In all these cases a variety of spatial patterns such as lamellas or dendrites is observed,

Figure 12 Left, skarn from San Leone, Sardinia. The light bands, 1 to 2 mm thick, consist of andraditic garnet (calcium and ferric-ion bearing). The dark bands, 5 to 8 mm thick, consist of magnetite and quartz. The white rectangle is 1 cm long. (Courtesy of B. Guy.) Right, orbicular diorite from Epoo, Finland. The concentric shells are alternately richer in biotite (dark) and plagioclase (light). The radius of the orbicule is 10 cm. (Courtesy of E. Merino.)

of characteristic scales larger by several orders of magnitude than any crystallographic length.

The recent promising application of ideas of nonequilibrium physics to geology is related to similar ideas. In numerous geological deposits spectacular regular mineralization structures are observed at a variety of space scales: metamorphic layers millimeters to meters thick, granites of centimeter-scale structure, agates with millimeter- to centimeter-wide bands, and others. Figure 12 shows two examples. The traditional interpretation attributes these structures to sequential phenomena, tracing the effect of successive environmental or climatic changes. It appears however that a more satisfactory interpretation would be to attribute them to symmetry-breaking transitions induced by nonequilibrium constraints. If this viewpoint is confirmed, the interpretation of numerous geological deposits will be deeply affected.

In addition to these examples, there are the problems of formation and propagation of defects and fractures, which involve, in one way or another, transitions mediated by the presence of interfaces.

Cooperative phenomena induced by electromagnetic fields: Electrical circuits, lasers, optical bistability

Electrical engineering and computer science make great use of cooperative phenomena arising in systems involving "active" elements such as Gunn diodes, resonant circuits coupled to a triode, or semiconductors. For example, the operation of almost all present-day computers involves circuits of bistable elements capable

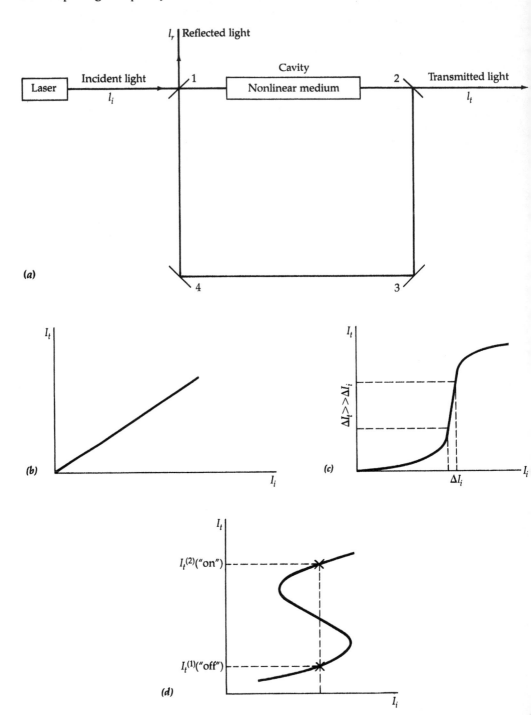

Figure 13 (a) Experimental setup to demonstrate optical bistability. (b) Transmitted versus incident light intensity in empty cavity. (c), (d) A cavity filled with an optically active medium can function as an optical transistor (c), or an optical switch (d).

of switching between two different states according to the values of certain parameters.

Since the 1960s a revolution has been going on in optics, a domain considered until then to be one of the most classical and well-established branches of physics. The generation of coherent light beams by the laser has stimulated impressive experimental and theoretical developments which have established that this phenomenon can also be viewed as a self-organization arising under nonequilibrium conditions. Moreover, optical bistability, the possibility of having two stable modes of operation of a resonant cavity, presents modern technology with the fascinating challenge of developing computers based on bistable electro-optical circuits, which would be more advantageous than electronic circuits because of shorter switching times and smaller consumption of energy.

Let us briefly outline the mechanism responsible for cooperative action and transitions in the particular case of optical bistability.

Suppose that a coherent field of electromagnetic energy having a sharply defined frequency in the visible range—that is, a beam of coherent light—is injected in a cavity, Figure 13(a). Such fields are generated quite easily by modern lasers. The cavity is adjusted to be in near-resonance with the incident field. Obviously, when the cavity is empty the intensity of the transmitted field I_t is proportional to the incident intensity I_i, as plotted in Figure 13(b). However, quite unexpected effects arise when the cavity is filled with a medium capable of absorbing the light and being in near-resonance with the incident beam. To begin with, the matter-light interaction is inherently nonlinear. For instance, the rate of change of the electric polarization of the sample in time depends on both the light intensity and the number of atoms that are in an excited state. More important, mirrors— 1, 2, 3, 4 in Figure 13(a)—direct part of the light that would otherwise escape back into the medium. This is equivalent to a positive feedback very similar to chemical autocatalysis. Under certain conditions this feedback is reflected by the behavior shown in Figure 13(c): a region of the transmission curve develops "differential gain," in the sense that the I_t versus I_i slope is greater than unity. The system then works as an *optical transistor*, since a low-intensity modulation imposed on the incident beam gets amplified at the output. Under other conditions the steady-state transmission curve becomes S-shaped as in Figure 13(d). We can then find a range of values of the incident intensity for which the system displays a bistable behavior. The two levels of transmission may be put in correspondence to the logical states "0" and "1," thus suggesting the possibility that such a device could be used as a switch in *optical memory* circuits.

1.7 AGAIN, BIOLOGICAL SYSTEMS

We cannot proceed in our survey of complex behaviors without discussing biological systems in a more systematic way, going beyond the few intuitive remarks made earlier.

Living beings are undoubtedly the most complex and organized objects found in nature, in view of their morphology and their functioning. As we have emphasized, they serve as prototypes from which physical sciences can get both motivation and inspiration for understanding complexity. They are literally historical structures, since they have the ability to preserve the memory of forms and functions acquired in the past, during the long period of biological evolution. Moreover, living systems function under conditions far from equilibrium. An organism as a whole continuously receives fluxes of energy (e.g., the solar influx used by plants for photosynthesis) and of matter (in the form of nutrients), which it transforms into quite different waste products evacuated to the environment. At the cellular level, strong inhomogeneities are also observed. For instance, the concentration of potassium ions, K^+, inside the neurons, the cells of the nervous system, is higher than in the outside environment while the opposite is true for sodium ions, Na^+. Such inequalities, which imply states of high nonequilibrium, are at the origin of processes such as the conduction of the nerve impulse, which play an important role in life. They are maintained by active transport and bioenergetic reactions like glycolysis or respiration.

Being convinced by now that ordinary physico-chemical systems can show complex behavior presenting many of the characteristics usually ascribed to life, it is legitimate to inquire whether some of the above features of biological systems can be attributed to transitions induced by nonequilibrium constraints and appropriate destabilizing mechanisms similar to chemical autocatalysis. This is probably one of the most fundamental questions that can be raised in science. No exhaustive answer can be claimed, but we can mention some examples in which the connection between physico-chemical self-organization and biological order is especially striking. The particular problem on which we focus here is the control of embryonic development.

Embryonic development is the sequence of events leading from a unique cell, the fertilized egg, to a complete organism. Nature provides us with an unlimited choice of illustrations of this process. Certainly one of the simplest cases is that of bacteria, whose development reduces to a sequence of cellular divisions. At the other end of the spectrum are advanced organisms like mammals whose development leads to a pluricellular body in which the cells form specialized tissues and organs that may comprise 10^{12} cells.

At present, it is out of the question to arrive at a detailed understanding of how such processes take place and, particularly, of how they are coordinated with the fantastic precision that allows each cell to fulfill its role at the right time and in the right place. Instead, we shall discuss living systems whose development is characterized by an intermediate level of complexity, like the amebas of the species *Dictyostelium discoideum*. Here development reduces essentially to a transition phenomenon marking the passage from the unicellular to the pluricellular stage of life. This is very similar to the phenomenon in the BZ reaction.

Figure 14 describes the life cycle of this species. In (*a*) the amebas are at the unicellular stage. They move in the surrounding medium; they feed on such nu-

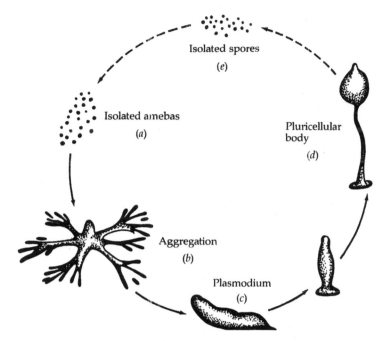

Isolated spores
(e)

Isolated amebas
(a)

Pluricellular
body
(d)

Aggregation
(b)

Plasmodium
(c)

Figure 14 Life cycle of the ameba *Dictyostelium discoideum*.

trients as bacteria and proliferate by cell division. Globally speaking they constitute a uniform system, inasmuch as their density (number of cells per square centimeter) is essentially constant. Suppose now that the amebas are subjected to starvation; in the laboratory this is induced deliberately, in nature it may happen because of less favorable ambient conditions. This is the analog of applying a constraint in a physical or chemical experiment. Interestingly, the individual cells do not die. Rather, they respond to the constraint by aggregating (b) toward a center of attraction. The initial homogeneity is broken; space becomes structured. The resulting multicellular body, the plasmodium (c), is capable of moving, presumably to seek more favorable conditions of temperature and moisture. After this migration the body differentiates (d) and gives rise to two kinds of cells, one of which constitutes the stalk and the other a fruiting body within which spores are formed. Eventually the spores are disseminated (e) in the ambient medium, and if the conditions are favorable they germinate to become amebas and the life cycle begins again.

Let us investigate the aggregation stage, shown in Figure 15, in more detail. The following phenomena are observed during this process. First, after starvation some of the cells begin to synthesize and release in the extracellular medium signals of a chemical substance known as cyclic adenosine monophosphate (cAMP). The synthesis and emission are periodic, just as in the chemical clock of the BZ system, with a well-defined period for given experimental conditions. The cAMP emitted by the "pioneer" cells diffuses in the extracellular medium and reaches

Figure 15 Concentric and spiral waves of aggregating cell populations of *Dictyostelium discoideum* on an agar surface. The bands of amebas moving toward the center appear bright and the stationary amebas dark.

the surface of the neighboring cells. Two types of events are then switched on. First, these cells perform an oriented movement called *chemotaxis* toward the regions of higher concentration of cAMP, that is, toward the pioneer cells. This motion gives rise to density patterns among the cells that look very much like the wave patterns in the BZ reagent (Figure 9). Second, the process of aggregation is accelerated by the ability of sensitized cells to amplify the signal and to *relay* it in the medium. This enables the organism to control a large territory and form a multicellular body comprising some 10^5 cells.

Thus, the response to the starvation constraint gives rise to a new level of organization resulting from the concerted behavior of a large number of cells and enabling the organism to respond flexibly to a hostile environment. What are the mechanisms mediating this transition? Let us first observe that the process of chemotaxis leads to an amplification of the heterogeneity formed initially, when the pioneer cells begin to emit pulses of cAMP. Because it enhances the density of cells near the emission center, chemotaxis contributes, by means of the phenomenon of relay, to an increased release of cAMP from this region, which further enhances the chemotactic movement of the other cells toward it. This constitutes what one usually calls a *feedback loop*, very similar to chemical autocatalysis or to the self-acceleration of an exothermic reaction encountered in Section 1.4.

As it turns out, a second feedback mechanism is present in *Dictyostelium discoideum* which operates at the subcellular level and is responsible for both the periodic emission of cAMP and the relay of the chemotactic signal. This mechanism is related to the synthesis of cAMP by the cell. cAMP arises from the transformation of another important cellular constituent, adenosine triphosphate, ATP, which is one of the principal carriers of energy within living cells thanks to its high-energy phosphate bond. The ATP → cAMP transformation is not spontaneous however; a catalyst is needed to accelerate it to a level compatible with vital requirements. In biological systems the task of catalysis is assumed by special molecules, the *enzymes*. These molecules usually contain several atoms and have a rather high molecular weight—hence the term *macromolecule* by which they are usually referred to. Some enzymes have a single active site which the reactants must hit in order to transform into products. But in many cases there are *cooperative enzymes*, which display several sites; some of the sites are catalytic and others are regulatory. When special effector molecules bind to the latter sites, the catalytic function is considerably affected. In some cases the molecules reacting with or produced from the catalytic site may also act as effector molecules. This will switch on a feedback loop, which will be positive (activation) if the result is the enhancement of the rate of catalysis, or negative (inhibition) otherwise.

The enzyme that catalyzes ATP → cAMP conversion is called adenylate cyclase and is fixed at the interior of the cell membrane. It interacts with a receptor fixed at the exterior phase of the membrane in a cooperative fashion, whose details are not completely elucidated. The cAMP produced diffuses in the extracellular

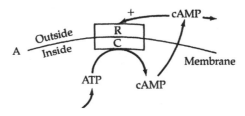

Figure 16 Mechanism of oscillatory synthesis of cAMP in slime mold *Dictyostelium discoideum*.

medium through the cell membrane and can bind to the receptor and activate it, as shown in Figure 16. In this way it enhances its own production, thereby giving rise to a feedback loop capable of amplifying signals and of inducing oscillatory behavior.

Simple as they may be, the processes underlying the development of *Dictyostelium dicoideum* are prototypical of a large class of more sophisticated developmental phenomena. It seems indeed that the breaking of spatial symmetry and the concomitant compartmentalization of cells explains many experimental data on insect morphogenesis. More generally, in developing tissues one frequently observes gradients of a variety of substances such as ions or metabolites. It is natural to conjecture that these gradients provide the tissue with a kind of "coordinate system" that conveys *positional information* to the individual cells, by means of which they can recognize their position with respect to their partners. It is therefore likely that transitions mediated by chemical substances and leading to symmetry breaking are one of the key features of life amenable to physico-chemical self-organization phenomena. This astounding idea was enunciated for the first time in 1952 by the British mathematician Alan Turing and ever since has been a constant source of inspiration for physicists and biologists alike.

1.8 COMPLEXITY AT THE PLANETARY AND COSMIC SCALE

We are still far from having defined complexity. In fact we may not achieve that goal before the end of the book, for complexity is one of those ideas whose definition is an integral part of the problems that it raises. What we do know by now, from the examples surveyed in preceding sections, is that one of the essential features of complex behavior is the ability to perform *transitions* between different states. Stated differently, complexity is concerned with systems in which evolution, and hence history, plays or has played an important role in the observed behavior.

How general are these systems? For a long time science held the static view that the world was a well-oiled machine working with the same implacable precision that characterized it as far back as one could conceive, and that would continue to characterize it in any foreseeable future. This view can hardly be sustained today. The biological revolution initiated by Darwin in the nineteenth century, and the revolution that has been going on in geophysics and cosmology since the 1950s, teach us that wherever we look in our environment at large we find relics of the past which, far from being mere fossils, are actively preparing the history yet to come. The earth's atmosphere as we know it today is a result of the development of life on the planet. Living beings are the results of a long evolution whose chemical, prebiotic stage was directly affected by conditions prevailing on earth some four billion years ago. The Greenland or Antarctic glaciers, the position of continents, the oceanic floor, our climate, all have been subjected to a long evolution characterized by a series of large-scale transitions. Last but not least, the universe as a whole is continuously evolving, and the microwave background radiation (equivalent to a black body radiating at 2.7 K) detected by astronomers is a compelling sign of a primordial transition that created the universe some 20 billion years ago.

We now begin to suspect how far transition phenomena similar to those taking place in a pan of liquid or in a chemical solution can lead. We cannot be exhaustive here, so we will select as a case study the problem of climatic change, to which we shall return in a later chapter. Further comments on the cosmic scale are presented in Section 1.9.

Since about 1960 an increasing number of observations have led to the conclusion that the earth's climate is capable of showing a very pronounced intrinsic variability. This fact, coming after the abnormally favorable and stable climatic period that humanity has known during the first part of our century, both astonishes and worries specialists, policy makers, and the public. For the first time humans realize the global, planetary character of the climatic system as well as the fact that their own activities can affect the functioning of the impressive climate machine.

Let us review a number of facts contributing to this new attitude, beginning with the problem of *long-term* changes. Most of the observations available are the results of the powerful technique of analysis of the isotopic composition of organic relics, giving direct information about paleotemperatures.

One of the first points that becomes obvious from such data is that the climatic conditions that prevailed in the last two or three hundred million years were extremely different from those of the present day. During this period, with the exception of the Quaternary era (our era, which began about two million years ago), there was practically no ice on the continents and the sea level was about 80 meters higher than at present. Climate was particularly mild, and the temperature differences between equatorial (25–30°C) and polar (8–10°C) regions were relatively weak.

It was during the Tertiary era, some 40 million years ago, that a sharper contrast between equatorial and polar temperatures began to develop. In the relatively short time of 100,000 years, the sea temperature south of New Zealand dropped by several degrees. This was probably the beginning of the Antarctic current, which reduces the exchange of heat between high and low latitudes and contributes to a further cooling of the masses of water "trapped" in this way near the polar regions. Once again, we see a feedback mechanism in action.

At the beginning of the Quaternary era this difference was sufficiently important to allow for the formation and maintenance of continental ice. In the northern hemisphere a series of *glaciations* took place in an intermittent fashion, sometimes pushing the glaciers as far as the middle latitudes. These climatic episodes present an average periodicity of about 100,000 years, though with considerable random looking variation, as shown in Figure 17.

The last advance of continental ice in the northern hemisphere attained its maximum some 18,000 years ago, and its relics are still with us. While the amount of continental ice today is about 30 million cubic kilometers, confined essentially to Antarctica and Greenland, there was at that time about 70 to 80 million cubic kilometers covering, in addition, much of North America and Northern Europe. Because of the huge quantities of water trapped in the glaciers, the sea level was some 120 meters lower than today. Since then a large part of the ice has melted, thus defining the coastlines and most of the other features of the present-day landscape.

Let us now discuss climatic changes at a shorter time scale. The last traces of continental ice caps (with the exception of Greenland and Antarctica) disappeared

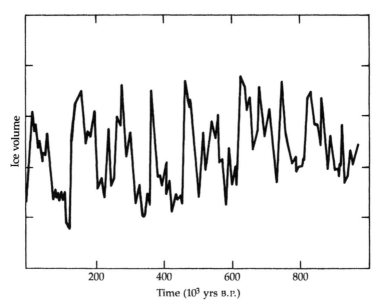

Figure 17 Variation of global ice volume during the last 1 million years, inferred from the isotope record of deep sea cores.

about 7000 years ago, and the period following this event is referred to as *climatic optimum*. An astonishing aspect of this era is the appearance of humid periods in the zone between Mauritania in West Africa (17°W) and Rajasthan in central India (78°E), which today is one of the most arid zones of the planet. Even in the center of the Sahara, which presently receives hardly five millimeters (0.2 inch) of precipitation per year, there were permanent rivers and an advanced agriculture, indicating an annual precipitation of at least 300–400 millimeters (12–16 inches). This favorable situation appears to have lasted until the beginning of the Iron Age, about 3500 years ago. The lifestyle of people in the East Mediterranean basin that the poems of Homer or the trowel of Schliemann reveal to us is reminiscent of a permanent spring, a generous nature, and a fertile land.

For about 3000 years a global cooling has occurred. However, this general trend has been interrupted by numerous fluctuations of about 1°C lasting for a century or more. A well-known example of warming fluctuation is the period A.D. 800–1200 during which the Vikings discovered Greenland (the green land) and probably North America. An opposite trend was observed between about 1550 and A.D. 1700. This "little ice age" caused food shortages and many other catastrophes in Europe and North America.

We now realize that the climate of the first half of our own century constituted an anomaly rather than a typical sample of the climate's long and tumultuous history. For several decades we have experienced relatively predictable weather, and a retreat of the northern hemisphere ice cap. Agriculture and food production have benefited considerably, and this has contributed to the increase of world population. On the other hand, this apparent permanence gave false ideas of what is or is not "normal" in climatology.

Since the mid-1970s there has been a return of climate variability. One example is the abnormally harsh winter that struck the eastern part of North America in 1976–1977 and the simultaneous prolonged drought in the western part of this continent. It seems that this phenomenon was related to a weakening, or even to a local "blocking," of atmospheric circulation as reflected, for instance, by large displacements of parts of the polar jet stream to the south (in the eastern part of the American continent) and to the north (in the western part of the continent).

Having seen that the climatic system has a long history of complex transition phenomena, let us survey some of the factors that may be at the origin of this complex behavior. As in the previous sections of this chapter, our discussion will remain qualitative. A more quantitative approach will be outlined in Chapter 6.

The principal factor controlling the earth's climate is undoubtedly the influx of solar energy. We name solar constant Q, the solar energy per unit time and surface received in the top of the atmosphere in a direction perpendicular to the sun's rays. Astrophysical calculations show that the sun's luminosity (just like the luminosity of the other stars of the so-called principal sequence) increases in time. It follows that some hundred million years ago the value of Q was lower by several percent from its present value. Still, we have seen that during this era the climate was very mild, and glaciations did not occur. This "paradox of faint early sun" shows that the earth's climatic system does not passively follow the variation of solar energy. In other words, internally generated processes—above all in this case the composition of the atmosphere, the clouds, and the vigorousness of atmospheric circulation—introduce a highly nontrivial modulation of external effects.

Internal processes also affect climatic response to a second external force, the variations in the earth's orbital characteristics. Astronomical calculations show that because of the perturbing action of the other bodies of the solar system, three major variations of the earth's orbit occur, relative to what is calculated by elementary mechanics. First, the earth's axis of rotation precesses like a spinning top, so that the direction in the sky toward which the axis points moves around a circle. Second, the tilt angle of the earth's axis of rotation and the direction perpendicular to the ecliptic (currently about 23.5°) varies by about ±1.5°. And third, the eccentricity of the earth's orbit is also subject to change. The characteristic periodicities of the above effects are, respectively, about 22,000, 41,000, and 100,000 years. The theory predicts that the first two effects change the seasonal distribution of solar energy, whereas the third affects the mean annual energy received by the earth. In all cases the direct effects are small. Still, the dominant periodicity of glaciation cycles is marked by them. Once again internal dynamics should play a highly nontrivial role by amplifying this weak external effect, thus triggering a major climatic change.

Let us give an example of such an amplifying mechanism. Consider the interaction between radiative processes and ice caps. It is well known that the reflectivity of ice—the *albedo*—is very high. Suppose that as a result of a perturbation whose origin need not be specified here, one of the ice caps advances slightly toward the equator. Because of the resulting increase of the global albedo, less energy would be absorbed by the earth, and this would cool the system further. And as a result the perturbed ice cap would advance even more, with a consequent supplementary lowering of temperature.We are therefore in the presence of a positive feedback loop very similar to chemical autocatalysis. Note that this is only one of the numerous complex interactions operating within the climatic system.

It should also be realized that the earth receives the major part of solar energy in the equatorial regions and sends back into space large parts of the lesser amount of energy received in the polar regions. In this way a meridional temperature gradient between the equator and poles is formed and sustained by the solar influx. This situation is similar to the thermally constrained system encountered in Section 1.3 and implies that the climatic system is maintained *far from thermodynamic equilibrium*.

Nonequilibrium, feedback, transition phenomena, evolution—once again we encounter the ingredients that were present in our previous examples of complex behavior, but operating here on the much more impressive planetary scale.

1.9 FORCES VERSUS CORRELATIONS—A SUMMING UP

A piece of iron at a temperature above 1044 K presents no detectable magnetic properties. But when we cool it below this "critical" temperature, the material becomes magnetizable. This is a typical example of an important class of natural phenomena known as *phase transitions*. Above the critical temperature the material is isotropic, in the sense that none of its observable properties is characterized by a preferred direction. But below this temperature in a well-defined region of the material, the so-called Weiss domain, a magnetization emerges which is a vector quantity pointing in a certain direction in space. The material becomes anisotropic and, in more technical terms, the *rotational symmetry* characterizing the state of no magnetization is broken.

There are many other examples. A liquid is a state of matter in which the molecules move in all possible directions and do not recognize each other over distances longer than a few hundred millionths of a centimeter [Figure 1 (a)]. It can therefore be regarded as a homogeneous material in which all points of space are equivalent. We now cool this system slowly and uniformly. Below a characteristic temperature (0°C for pure water at ordinary pressure), we obtain a crystal lattice, a new, solid phase of matter. Its various properties, for example, the density, are no longer identical as we move along a certain direction in space; in other words, the *translational symmetry* characterizing the liquid is broken.

In both cases the breaking of symmetry accompanies the appearance of new properties that prompt us to characterize the material as *ordered*. For instance, in the crystal lattice the molecules perform small vibrations around regularly arranged spatial positions which, depending on the case, may lie at the vertices of a cube, or the vertices of a regular hexagonal prism and the centers of their hexagonal bases, and so on.

What is happening here? Did we simply forget to include these phenomena in our long list of transitions to self-organization and complex behavior?

In the absence of interactions, the molecules of a material move freely in all directions, and this motion is entirely characterized by their individual kinetic energies $1/2(m_i v_i^2)$ (m_i, v_i being the molecular masses and velocities), which remain invariant in time. We know, however, that in any physical system the molecules interact, essentially by forces of electromagnetic origin. A typical example are the van der Waals forces prevailing in an electrically neutral fluid. These are *short-range* forces whose intensity drops abruptly to extremely small values beyond an interparticle separation of a few molecular diameters, as shown in Figure 18.

When the system is dilute or when temperature is very high, intermolecular forces are not very efficient: kinetic energy dominates, and the material behaves in a disordered fashion. But when we lower the temperature (or compress the system), the roles of kinetic energy and of interaction potential tend to be reversed. Eventually the system is dominated by the interactions and adopts a configuration

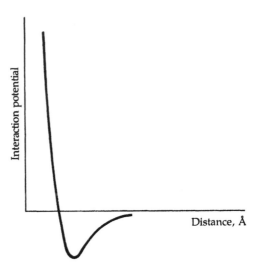

Figure 18 A typical form of intermolecular interaction potential in a neutral gas. The distance corresponding to the minimum of the curve is a few angstroms and the value of the potential energy at this point is about $k_B T_c$, where k_B is Boltzmann's constant and T_c is the critical temperature of liquid-vapor transition.

in which the corresponding potential energy is as small as possible. Being mediated entirely by the intermolecular interactions, this configuration displays a characteristic length—like the distance constant in crystal lattice structure—that is microscopic (a few angstroms), in keeping with the range of these interactions. Moreover, the behavior of the material is time-independent if the environment itself is stationary. In the light of these remarks we can say that the order associated with a phase transition leads to "fossil" objects.

These properties must be contrasted with those characterizing the transition phenomena surveyed in the preceding sections of this chapter. Indeed, the space scales characterizing the states emerging beyond these latter transitions are *macroscopic*, comparable to or larger than the characteristic scales of biological systems. Moreover, in addition to spatial patterns a variety of time-dependent phenomena can arise and be sustained whose characteristic scales are again macroscopic.

The reason behind this basic difference is that a new organizing factor not amenable to intermolecular interactions is now at work: the *nonequilibrium constraint*. Because it encompasses parts of the system containing huge numbers of molecules, new phenomena associated with states of matter enjoying long-range correlations are born. It is instructive to realize that in a dendritic structure like a snowflake these two kinds of order are superimposed and can thus easily be distinguished and opposed. Indeed, the underlying crystal lattice has little to do with the emergence of dendrites or their size and spacing, which are orders of magnitude larger than any crystallographic length. In short, the complex behavior we are interested in here can be thought of as a phase transition of a new type, one in which the lowering of temperature is replaced by the application and progressive increase of suitable nonequilibrium constraints. For this reason the term "nonequilibrium phase transition" is sometimes used in connection with self-organization phenomena leading to complex behavior.

Force-mediated symmetry-breaking phenomena that are similar in some respects to phase transitions also appear at the level of fundamental interactions. One of the most striking examples is the broken symmetry between matter and antimatter. As is well known, quantum relativistic physics associates to each charged particle an antiparticle of the same mass but opposite charge: there are electrons and antielectrons (positrons), protons and antiprotons, and so on. Experimentally, antiparticles have been observed under special conditions. However, although the basic equations of physics are symmetric in matter and antimatter, the proportion of antimatter in the observable universe is negligible with respect to matter. How did this asymmetry come about?

One possible answer comes from the grand unified theories of the strong, electromagnetic, and weak interactions of elementary particles, which are believed to describe the behavior of matter shortly after the start of our universe. In particular, it appears possible that matter underwent one or perhaps several transitions at "critical temperatures" as it cooled from the initial exceedingly hot state, owing to the expansion of the universe. According to the grand unified theories, matter above the first transition temperature was in a very symmetric state. Quarks, the constituents of protons and neutrons, and leptons—including electrons, neutrinos, and their antiparticles—all behaved identically. Below the transition point, however, the differences were made manifest through symmetry breaking. Eventually these differentiated particles became the raw material that makes up stars, planets, and living beings. In this sense, therefore, a transition mediated by the above three exceedingly short-range fundamental interactions has reached a macroscopic scale encompassing the present universe, which appears to be a "relic" of an event produced in a remote past. There is no contradiction, however, between this statement and our previous effort to oppose ordinary phase transitions and nonequilibrium phase transitions on the basis of their widely different ranges. Indeed, the history of the early universe, and in particular its expansion, is marked by the passage from a state of thermal equilibrium to a state in which the equilibrium between different constituents of matter as well as between matter and radiation is broken. In such a nonequilibrium world large-scale symmetry-breaking transitions of the kind mentioned above can take place. In this respect therefore, differentiated matter as we observe it today can be viewed as the outcome of a primordial nonequilibrium.

The existence of such primordial nonequilibrium processes has deep consequences for our view of the evolution of the universe. In Chapter 2 we will introduce the celebrated second law of thermodynamics, dealing with the increase of entropy. It may be interesting to recall that in his original formulation in 1865, Rudolf Clausius enunciated this principle in cosmological terms: "The energy of the universe is constant; the entropy of the universe increases toward a maximum." Thus, the thermodynamic evolution of the universe would go from low entropy to high entropy. This led to the idea of "thermal death," which should be our inexorable fate. However, recent developments in cosmology are challenging this view. In 1965, exactly one century after Clausius formulated the second law, Penzias and Wilson discovered the residual black-body radiation, showing that our observable universe is full of photons. In fact, there is only one baryon for 10^8 or 10^9 photons. So, essentially, matter is floating in a bath of photons. This means that by far the most important component of entropy is constituted by these photons, which were produced during the early stages of our universe. In this sense, we could say that thermal death is no more in our future, but lies behind us, in the first stage of our universe.

To describe the irreversible processes that have led from such an initial situation to a universe full of structure and diversity we need to take into account gravitation, the fourth fundamental force of nature, which is well known for its long-range character. Modern gravitation theory was developed after the formulation of Einstein's theory of general relativity. Let us try to give the flavor of the relation that may exist between gravitation, general relativity, and irreversible processes, referring for more details to Appendix 5.

As is well known, the world, as described in newtonian theory, presented an intrinsic duality: on one side, space-time; on the other, matter, moving under the influence of forces. The great idea of Einstein was to go beyond this duality, and to relate space-time to matter. General relativity is the basis of all models of cosmological evolution.

However, there are some difficulties. Einstein's equations describe an evolution in which matter and entropy are conserved. Therefore, if we go back to the far distant past, we arrive at a singularity for which all matter and entropy are concentrated in a single point. This is the famous Big Bang singularity. A. Wheeler has called this singularity "the greatest crisis of physics." Indeed, beyond it, the laws of physics have no meaning. Today, many attempts are made to avoid this singularity, replacing it by a phase transition of some kind. What we want to emphasize here is that during this primordial phase, irreversible processes are likely to play an essential role. In fact, in our view the very appearance of matter out of vacuum would be the basic irreversible process at the start of our universe.

The
Vocabulary of
Complexity

*I*n Chapter 1 we adopted the naturalist's point of view; we tried to get an intuitive feeling of complexity by observing transition phenomena that lead to self-organization in various contexts. In doing so we encountered a number of new ideas, and we realized that they recurred continuously throughout our examples. Now we must take a more systematic, deeper look at these ideas in order to establish the *vocabulary of complexity*. The development in the present chapter lays the groundwork for Chapters 3 through 5, which present the tools needed to deal with our ideas about complexity in a creative way.

2.1 CONSERVATIVE SYSTEMS

Ever since the birth of abstract thinking, which marked the beginning of science and modern civilization, the concept of *conservation* has played a key role. The great Greek thinker Thales, traditionally considered to be the earliest philosopher, held the view that the multiple facets of nature could be reduced to a single fundamental substance, water. In modern terms he viewed the universe as a *conservative system* in which, despite continuous interchange among its different parts, there existed a primordial element that remained untouched by change. Similar thoughts must have deeply influenced the young Plato, whose theory of Ideas is essentially a brilliant quest for permanence at a more abstract level.

Two thousand years later we encounter another great period of human history, that of Isaac Newton and Gottfried Wilhelm Leibniz. With the advent of classical mechanics at that time, the concept of conservation was expanded and developed in great detail. Conservation no longer relied on some uncontrollable axiom, rather it was the consequence of a set of laws that provided a quantitative explanation of observations on the motion of celestial bodies and the motion of objects under the effect of terrestrial gravity. Indeed, Newton's second law relating force to acceleration, and third law relating "action" to "reaction," imply that in a system of interacting massive points free of any external influence, three quantities remain invariant in time:

1. Total energy
2. Total translational momentum
3. Total angular momentum

The forces of interaction between two bodies are shown in Figure 19, and the invariant quantities just listed are discussed below.

1. The total energy,

$$E = \text{Kinetic energy} + \text{potential energy}$$
$$= \sum_i \tfrac{1}{2} m_i v_i^2 + \tfrac{1}{2} \sum_{i \neq j} V_{ij} \tag{2.1a}$$

in which m_i and v_i are respectively the mass and velocity of material point i, and V_{ij} is the interaction potential between i and j, related to the force F_{ij} acting on i due to the presence of j through

$$\text{Work of force} = \text{Change in potential energy}$$

For a force that depends only on the distance $r \equiv r_i - r_j$, this reduces to

$$F_{ij}(r) = -\nabla_r V_{ij}(r) \tag{2.1b}$$

∇_r being the derivative of the potential with respect to the distance.

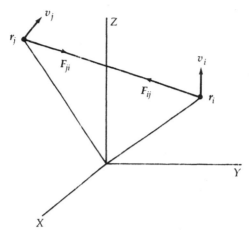

Figure 19 Central forces of interaction between two particles located at r_i, r_j having velocities v_i, v_j. By Newton's third law, $F_{ij} = -F_{ji}$.

2. The total translational momentum,

$$P = \sum_i p_i = \sum_i m_i v_i \tag{2.2}$$

3. The total angular momentum,

$$L = \sum_i l_i = \sum_i m_i r_i \times v_i \tag{2.3}$$

in which l_i describes the rotational motion of the ith material point around a certain point in space chosen for convenience as the origin of coordinates.

All these conservation laws are intimately related to a key property of the basic equations of mechanics, namely, that the quantity related dynamically to the force (usually a function of the coordinates r_i) is the acceleration, the *second derivative* of position (not the velocity, as Aristotle thought):

$$m_i \frac{d^2 r_i}{dt^2} = \sum_j F_{ij}(r_i - r_j) \tag{2.4}$$

Hence, an observer recording the motion toward the past ($t' = -t$) will detect exactly the same properties as the usual terrestrial observer who makes measurements for "positive," future-directed times. Alternatively, realizing that $d^2 r/dt^2 = dv/dt$ and $v = dr/dt$, we can see that the transformation $t' = -t$ induces the further transformation $v' = -v$, leaving the acceleration invariant. The material points follow the same trajectories, but the direction of flow is reversed. There is nothing special in such a change, since we could have given them initial velocities sufficient for sustaining flow in this latter direction for positive times. In other words, nothing in the equations of motion allows us to differentiate between direct and reversed motion.

Classical mechanics thus appears to be the science of both conservative and time-reversible systems. An elegant approach is based on William Hamilton's formulation, in which the structure of the equations of motion is related directly to energy—the most important of the three invariant quantities listed earlier. Specifically, introducing the hamiltonian function, H,

$$H = H(p, q) = E = \text{constant}$$

in which p, q generalize the notions of momentum ($p = mv$) and position ($q = r$) of the newtonian formulation, we have:

$$H = \sum_i \frac{p_i^2}{2m_i} + \frac{1}{2} \sum_{ij} V_{ij}(q_i - q_j) \tag{2.5a}$$

It is easy to check that Eq. (2.4) and the relation $v = dr/dt$ are equivalent to:

$$\frac{d}{dt} q_i = \frac{\partial H}{\partial p_i}$$

$$\frac{d}{dt} p_i = -\frac{\partial H}{\partial q_i} \tag{2.5b}$$

Today, hamiltonian dynamics is an important part of science. In Chapter 3 we will briefly describe the revolutionary developments that have marked it since the 1950s. Here we identify a few examples of conservative systems.

One of the most familiar conservative systems is the pendulum. If we denote the pendulum length by l, the acceleration of gravity by g, and the angle from the downward vertical by θ and agree to measure energies using the position of the mass on this vertical as a reference, we can write the total energy—the hamiltonian—in the form

$$H = \text{Kinetic energy} + \text{potential energy}$$

$$= \tfrac{1}{2}mv^2 + mgl(1 - \cos \theta)$$

Since v is tangent to the trajectory of the mass, $v = l(d\theta/dt)$, and thus

$$H = ml\left[\tfrac{1}{2}l\left(\frac{d\theta}{dt}\right)^2 + g(1 - \cos \theta) \right] = \text{constant} \tag{2.6a}$$

As discussed in more detail in Chapter 3, it is very useful to visualize the evolution of a system in an abstract space known as *phase space*. In phase space the coordinates are the positions and momenta appearing in Hamilton's equations, Eqs. (2.5b). In the case of the pendulum, $q = \theta$, $p = ml(d\theta/dt) \equiv ml\dot{\theta}$, the phase space is the plane $(\dot{\theta}, \theta)$. The instantaneous state of the pendulum is thus a point in this space, referred to as the *representative point*. As time varies, both θ and $\dot{\theta}$ take uniquely defined values; in other words, the representative point describes a one-dimensional curve—the phase space trajectory. The tangent to this curve is the phase space velocity. These trajectories and velocities are represented in Figure 20. They should not be confused with the more familiar trajectory and velocity of a body moving in physical space which, despite their usefulness, provide a less detailed description of the motion.

A particular case of a phase trajectory is the *position of equilibrium*, in which both the velocity and acceleration are zero. This is a quite degenerate trajectory, since it is represented by a single point in phase space. In the pendulum there are two such equilibria, shown in Figure 20 lying on the downward vertical ($\theta = 0$), and on the upward vertical ($\theta = \pm\pi$). As we shall see, they have a quite different status.

We know from observation and from elementary mechanics that for small initial velocities and deviations from the downward vertical, the pendulum makes sustained oscillations in time in the form of small vibrations. This means that both θ and $\dot{\theta}$ run over the same range of values after each period. In the phase space representation this will be described by a trajectory that is a *closed curve* (*a* or *a'* in Figure 20) along which the energy, Eq. (2.6a), will be conserved.

Another familiar conservative system is the harmonic oscillator, which models the frictionless motion of a particle of mass m attached to one end of a horizontal, perfectly elastic spring whose other end is rigidly fixed. The force exerted on the particle is proportional to the deviation, x, of the length of the spring from its

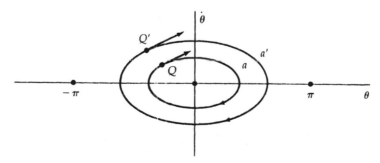

Figure 20 Phase space $(\dot{\theta}, \theta)$ of the pendulum $(-\pi \leqslant \theta \leqslant \pi)$. $\theta = \dot{\theta} = 0$: stable equilibrium; $\dot{\theta} = 0$, $\theta = \pm\pi$: unstable equilibrium, *a*, *a'*: periodic trajectories corresponding to two different values of energy; *Q, Q'*: representative points on these trajectories.

equilibrium value. Denoting by k the coefficient of proportionality, we can write the hamiltonian of the system as the sum of the kinetic energy of the particle and the elastic energy of the spring:

$$H = \tfrac{1}{2}m\dot{x}^2 + \tfrac{1}{2}kx^2 \tag{2.6b}$$

A still different class of conservative systems arises in the two-body problem, the motion of two bodies interacting by gravitational or any other force that depends only on the absolute value of the distance between the bodies. However, complications arise quickly in mechanics. With two pendulums coupled through a spring, a single pendulum forced periodically from outside, or three interacting masses (the celebrated three-body problem), we arrive at the frontier of present knowledge. We will discuss these problems in some detail in Chapter 3.

We should not conclude from the above discussion that conservative systems are confined to classical mechanics. The propagation of light in a vacuum, described by Maxwell's well-known equations, provides an important example of conservative systems in electromagnetism. Conservative systems also arise in quantum mechanics, in problems that deal with the properties of simple atoms and molecules in the absence of external fields.

2.2 DISSIPATIVE SYSTEMS

In addition to conservative systems as described by classical dynamics, we must consider systems that give rise to irreversible processes. A very simple example is provided by systems exhibiting friction.

The importance of friction, a particular form of dissipation, was perceived long before classical dynamics was formulated. When Aristotle assumed that terrestrial dynamical systems in general tend to a state of rest, he was expressing the idea that some sort of "friction" would slow down the motion of the system components. In contrast, the classical principle of inertia, which states that acceleration, not velocity, is the basic mechanical quantity, corresponds to a model in which friction is neglected.

Starting with the work of Jean Baptiste Fourier and Rudolf Clausius, interest in dissipative systems that give rise to irreversible processes became increasingly widespread during the nineteenth century. This was quite natural in a period that witnessed the industrial revolution. However, in this context dissipation appeared to be associated with degradation, the exhaustion of available energy.

It is interesting that among the great ancient philosophers, Plato was acutely aware that both permanence and change must be integral parts of reality. But in the nineteenth century a conflict appeared. In physics, irreversibility and dissipation were interpreted as degradation, while among natural scientists biological evolution, which is obviously an irreversible process, was associated with increasing complexity. Owing perhaps to its technological implications, fluid mechanics was historically the first field in which the crucial role of dissipative processes was fully realized. As the idea of the molecular constitution of matter gradually established itself in science, so too did chemical kinetics, brownian motion, and various kinds of transport phenomena. Today scientists realize that dissipative systems constitute a very large and important class of natural systems. Most of our examples in Chapter 1 dealt with phenomena taking place in such systems, not the least of which is life itself.

The most conspicuous difference between conservative and dissipative systems appears in a *macroscopic description* of the latter, which uses collective variables such as temperature, concentration, pressure, and convection velocity to define the instantaneous state. When the equations of evolution of these variables are studied an important general feature emerges, namely, that the structure of the equations is not invariant under time reversal, contrary to Eqs. (2.4) and (2.5b). We therefore expect an irreversible course of events.

A striking illustration of this situation is provided by a chemical reaction. Consider the process described by Eq. (1.2a). The rate at which particles of species A are consumed is proportional to the frequency of encounters of molecules of A and B—which, if the system is dilute, is merely proportional to the product of their concentrations, c. Thus,

$$\frac{dc_A}{dt} = -kc_A c_B \tag{2.7a}$$

Clearly, if we reverse time, $t' = -t$, and denote by c_A', c_B' the values of the concentrations as functions of t', Eq. (2.7a) becomes

$$\frac{dc_A'}{dt'} = kc_A' c_B'$$

and describes a process in which c_A would be produced instead of being consumed. This is certainly not equivalent to the phenomenon described by Eq. (2.7a).

Further examples of dissipative processes are heat conduction and diffusion. Experiment shows that when a slight inhomogeneity appears within a uniform fluid it will spread out and eventually disappear. The same one-way evolution will be observed when a slight temperature variation is imposed locally and momentarily in an isothermal fluid. The following equations provide a quantitative description of these phenomena that is in excellent agreement with experiment:

$$\frac{\partial c}{\partial t} = D\nabla^2 c \qquad D > 0 \qquad \text{(Fick's equation)} \tag{2.7b}$$

$$\frac{\partial T}{\partial t} = \kappa\nabla^2 T \qquad \kappa > 0 \qquad \text{(Fourier's equation)} \tag{2.7c}$$

Here c denotes the concentration of a certain substance dissolved in the fluid, T is the temperature, D is the mass diffusivity, and κ is the heat diffusivity coefficient. Again, if we reverse time, we obtain the completely different laws

$$\frac{\partial c'}{\partial t'} = -D\nabla^2 c'$$

$$\frac{\partial T'}{\partial t'} = -\kappa\nabla^2 T'$$

describing a situation in which an initial concentration or temperature disturbance would be amplified rather than damped.

Both the concentration and the temperature variables are examples of so-called *even variables*, whose sign does not change upon time reversal. In contrast, the momentum of a particle and the convection velocity of a fluid are *odd variables*, since they are ultimately expressed as time derivatives of positionlike variables and change their sign with time reversal. This leads us to the following general property of the evolution equation of a dissipative system. Let $X_1, \ldots, X_n$ denote a complete set of macroscopic variables of such a system. Their time evolution will take the form

$$\frac{\partial X_i}{\partial t} = F_i(X_1, \ldots, X_n, r, t, \ldots) \tag{2.8}$$

in which F_i may be complicated functions of the X's and their space derivatives, as well as explicit functions of space r and time t. Then, in a dissipative system, if we perform a time reversal, $t' = -t$, at least one of the rate functions F_i corresponding to an even variable X_i must contain a part that remains invariant, whereas the rate function F_i corresponding to an odd variable X_i must contain a part that changes sign under time reversal. An example of rate functions of the even-variable class is provided by the righthand sides of Eqs. (2.7a) to (2.7c); an example of the odd-variable class is the contribution of viscosity to the balance equation of momentum of a fluid undergoing convection.

A crucial additional step in the distinction between conservative and dissipative systems will be made in Section 2.5, where the second law of thermodynamics and its multiple implications are discussed.

Just as for conservative systems, a convenient phase space can be introduced for dissipative systems. It is spanned by the ensemble of variables present and thus becomes a space of an infinite number of dimensions in the case of a continuous medium in which the various properties are distributed in space [Eqs. (2.7b), (2.7c)]. Phase space is thus much easier to handle when the number of variables is discrete and, particularly, finite or small. In the example of Eq. (2.7a), the phase space reduces to a line and the phase trajectory necessarily follows this line, as shown in Figure 21(a). A less trivial example is provided by the chemical reaction scheme

$$A \xrightarrow{k_1} X \xrightarrow{k_2} Y \xrightarrow{k_3} F \tag{2.9a}$$

with the rate equations

$$\frac{dc_x}{dt} = k_1 c_A - k_2 c_X$$

$$\frac{dc_y}{dt} = k_2 c_X - k_3 c_Y \tag{2.9b}$$

The phase trajectories are shown in Figure 21(b).

It is instructive to realize that certain dissipative systems can be written in a form reminiscent of conservative systems and placed in a hamiltonian framework. An example is the celebrated Lotka-Volterra mechanism,

$$A + X \xrightarrow{k_1} 2X$$

$$X + Y \xrightarrow{k_2} 2Y \tag{2.10}$$

$$Y \xrightarrow{k_3} D$$

in which a nontrivial constant of motion exists, playing the role of a hamiltonian. Still, in spite of the apparently conservative character, there is no time reversal invariance since both X and Y represent positive variables. It is therefore meaningless to ascribe to the variables properties similar to those of momentum in classical mechanics, properties that are necessary to ensure such an invariance.

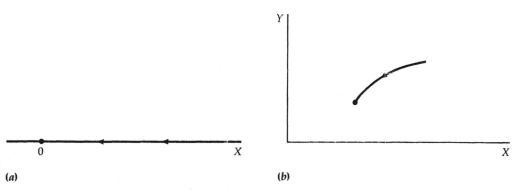

(a)   (b)

Figure 21 Phase space representation of a dissipative system. (a) One-variable system of Eq. (2.7a). (b) Two-variable system of Eqs. (2.9b).

A basic question not discussed here is how dissipative systems are related to conservative systems, and whether a passage from one to the other can be found. The major part of Chapter 5 is devoted to this problem.

2.3 *MECHANICAL AND THERMODYNAMIC EQUILIBRIUM. NONEQUILIBRIUM CONSTRAINTS*

In mechanics, equilibrium is a particular state in which both the velocities and the accelerations of all the material points of a system are equal to zero. We encountered this type of "state of rest" in Sections 1.1, 1.4, and 2.1. By definition the net balance of forces acting on each point is zero at each moment. If this balance is disturbed, equilibrium will be broken. This is what happens when a piece of metal fractures under the effect of load.

Certain familiar mechanical systems function in such a way that equilibrium is never fulfilled. For instance, the orbit of the earth around the sun is a manifestation of a permanent deviation from mechanical equilibrium. The latter would be possible only for an infinite separation of the two bodies.

The notion of *thermodynamic equilibrium*, encountered repeatedly in Sections 1.3, 1.4, 1.6, and 1.7, is sharply different. Contrary to mechanical equilibrium, the molecules constituting the system are subjected to forces that are not balanced and move continuously in all possible directions unless the temperature becomes very low. "Equilibrium" refers here to some collective properties characterizing the system as a whole, such as temperature, pressure, or the concentration of a chemical constituent. Let us outline a general formulation of this concept, of which the examples of Chapter 1 will turn out to be particular cases.

Consider a system embedded in a certain environment as represented in Figure 22. We characterize these entities by a set of properties, denoted X_i and X_{ie},

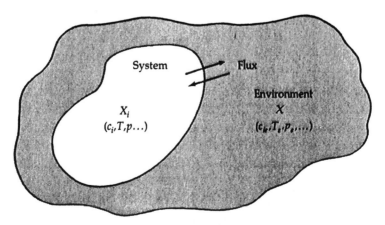

Figure 22 Schematic representation of an open system.

whose dynamic role resides primarily in their relation to the *exchanges* between the system and the environment. For instance, if the system is contained in a vessel whose walls are perfectly rigid, permeable to heat but impermeable to matter, one of these quantities will be identical to the *temperature, T* and will control the exchange of energy in the form of heat between the system and its environment.

We say that the system is in thermodynamic equilibrium if it is completely identified with its environment, that is, if the properties constituting the sets X_i and X_{ie} have identical values. In the previous example, thermodynamic equilibrium between the system and its surroundings is tantamount to $T = T_e$ at all times and at all points in space. But because the walls of the vessel are impermeable to matter, system and environment can remain highly differentiated in their chemical composition, *c*. If the walls become permeable to certain chemical substances *i*, thermodynamic equilibrium will prevail when the system and the environment become indistinguishable as far as those chemicals are concerned. In simple cases this means that the corresponding composition variables will satisfy the equality $c_i = c_{ie}$, but more generally equilibrium will be characterized by the equality of a quantity known as the *chemical potential*, $\mu_i = \mu_{ie}$. Similarly, if the walls of the vessel are not rigid, the system can exchange mechanical energy with its environment. Equilibrium will then also imply the equality of *pressures*, $p = p_e$. Further variables characterizing thermodynamic equilibrium may arise in the presence of external forces, for example, an electric field applied to a polarizable medium, a stress applied to a deformable material such as a plastic solid, and so on.

The case of *isolated systems*, which can exchange neither energy nor matter with their environment, deserves special mention. By definition, we cannot say that at equilibrium an isolated system is identified with its environment. Even so, much of the above discussion is valid since a small part of an isolated system is subjected to an environment constituted by the remaining parts of the system. As a result, at equilibrium the different parts of the system will become indistinguishable.

According to the above definitions, equilibrium is automatically a stationary state; mathematically, $\partial X_i / \partial t = 0$: the properties do not vary with time. It is, however, a stationary state of a special type since, because they are identical in the properties X_i, the system and the environment have nothing to exchange. For instance, in an equilibrium conditioned by equality of temperature, chemical composition, and pressure, there will be no exchange of thermal or mechanical energy or matter. We express this by saying that there are no net *fluxes* across the system:

$$J_i^{eq} = 0 \qquad\qquad (2.11)$$

This equality does not imply that equilibrium is a state of complete rigidity. Rather, it should be understood in a dynamical sense, as explained in the beginning of Section 1.4 with regard to chemical reactions. Specifically, for each process inducing a variation of J_i (say a slight increase of T from T_e in a small part of the system), there will be an inverse process contributing a variation of J_i in an exactly opposite direction. This basic property of *detailed balance* is a manifestation of the time reversibility of the elementary processes going on in the system, which obey the laws of conservative systems surveyed in the first section of this chapter. Detailed balance is responsible for most of the strong properties of equilibrium, such as uniqueness and stability. These properties will be discussed further in Sections 2.4 to 2.6.

It is now an easy matter to extend our discussion to *nonequilibrium states*. Such states are associated with nonvanishing fluxes between system and environment and therefore with differences between some of the state variables X_i and X_{ie}. They can be *transient*, in the sense that they can arise momentarily because of some initial condition and eventually relax as the system equilibrates with its environment. But they can also be permanent if we establish and maintain appropriate conditions, which we refer to as *constraints*. Thus, a temperature difference applied between two sections of a slab, or the pumping in of an initial reactant and the simultaneous removal of a final product in a chemical reaction system, will result in nonequilibrium situations in which the system is never allowed to identify itself with its environment. We should not conclude from these examples that nonequilibrium is an artificially imposed condition. As pointed out in Section 1.8, we see nonequilibrium states in much of our natural environment—for example, the state of the biosphere, which is subjected to an energy flux that arises from the balance of radiation between the sun and the earth.

Because of the continuous or temporary action of a constraint, in a nonequilibrium state detailed balance will not hold. As a result, a regime of nonequilibrium becomes susceptible to change. Localized small attempts to deviate from it are not necessarily obliterated by an instantaneously developed counteraction but, rather, can be accepted and even amplified by the system, thus becoming sources of innovation and diversification. As we shall see, this accommodation is at the origin of the ability of nonequilibrium systems to undergo bifurcations to new states bearing no resemblance whatsoever to the state of equilibrium.

2.4 NONLINEARITY AND FEEDBACKS

In the previous section we gave an operational definition of equilibrium states and nonequilibrium constraints. We now adopt a dynamical point of view, with the aim of establishing a relation between these concepts and the laws governing the evolution of the state variables. We have already established the connection for mechanical equilibrium in our discussion at the end of Section 2.1. Here we focus on dissipative systems, although many of the concepts that will be introduced could easily be applied to conservative systems.

The mathematical expressions describing how a set of state variables $\{X_i\}$ evolves in time have already been given in Eqs. (2.8). We shall rewrite these equations by introducing the new idea that arose in Section 2.3, namely, that the evolution is conditioned by appropriate nonequilibrium constraints. More generally, we want to express the idea that evolution is influenced by the variation of some parameters present in the problem that can be modified by the external world. We call these entities *control parameters*, denote them by λ, and write Eqs. (2.8) in the form

$$\frac{\partial X_i}{\partial t} = F_i(\{X_j\}, \lambda) \tag{2.12}$$

In a dissipative system these equations must be compatible with the conditions discussed at the end of Section 2.2. Equations (2.9b) provide a typical example of such dissipative *evolution laws*.

In view of the diversity of natural phenomena, we expect that the structure of the functions $\{F_i\}$ will depend in a very specific way on the system considered and on the type of process going on in this system. Indeed, at first sight there is absolutely no reason why the flow of the fluid contained between the two plates in the Bénard problem should be related in any sense to the aggregation of the amebas of *Dictyostelium discoideum* or to the glaciation of the Quaternary era. However, certain basic features can be sorted out of this apparently bewildering variety, and they will help us tackle complex phenomena in a systematic fashion.

One of these features is that, whatever the form of F_i, in the absence of constraints Eqs. (2.12) must reproduce the state of equilibrium. Since equilibrium is a steady state, this will amount to

$$F_i(\{X_{j,eq}\}, \lambda_{eq}) = 0 \tag{2.13a}$$

More generally, for a nonequilibrium steady state,

$$F_i(\{X_{j,s}\}, \lambda) = 0 \tag{2.13b}$$

These relations impose certain restrictions. For instance, the evolution laws must ensure that positive values are attained for temperature or chemical concentrations that come up as solutions, or that detailed balance is attained. This is an important point, for it shows that the analysis of physical systems cannot be reduced to a mathematical game. In many respects physical systems may be regarded as highly atypical or nongeneric from the mathematical point of view. Think, for instance, of conservative hamiltonian systems that are built on the extremely stringent condition that a particular combination of their positions and velocities—the energy—remains an invariant of the motion.

Granted that this specificity of physical systems is respected, now what can be said about the possible solutions of Eqs. (2.13a) or (2.13b)? It is here that the concept of nonlinearity begins to play a prominent role.

Let us take an example. Let X be the unique state variable, k a parameter, and let λ represent the applied constraint. We can easily imagine a mechanism such as $A \rightleftarrows X \rightleftarrows D$ in which X evolves according to $dX/dt = \lambda - kX$, yielding a stationary state value given by

$$\lambda - kX_s = 0$$

or

$$X_s = \frac{\lambda}{k} \tag{2.14}$$

Figure 2.3(a) represents this solution. By following the way the stationary state value X_s varies in terms of the constraint λ (which may denote, in particular, the distance from equilibrium), we obtain a straight line passing from the origin. This is a typical example of a *linear law*. For each value λ of the constraint there corresponds one and only one stationary state, X_s. Moreover, since two points suffice to define a straight line, by experimentally determining a pair of points $(\lambda_1, X_{1s}; \lambda_2, X_{2s})$, we can predict in a perfectly unequivocal manner the stationary state for any value of λ. In other words, in our linear system the behavior is bound to be qualitatively similar to that in equilibrium, even in the presence of strongly nonequilibrium constraints.

Let us now look at the situation depicted in Figure 2.3(b), which describes a *nonlinear law* linking the steady state value X_s to the control parameter λ. The curve drawn in this figure is only an example, since there is an unlimited number of possible forms describing nonlinear dependencies. When λ is less than λ_1 or

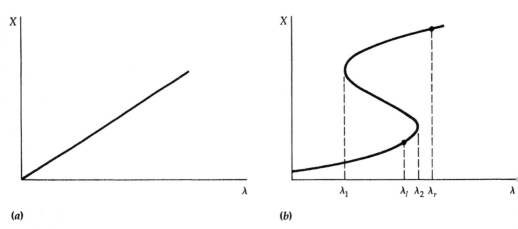

(a)   (b)

Figure 2.3 Effect of a parameter λ on a state variable X. (a) Linear law. (b) Nonlinear law. Notice the sensitivity of the response of X with respect to changes of λ in the vicinity of the values λ_1, λ_2 in case (b).

larger than λ_2, the situation is as in Figure 23(a). But for λ between λ_1 and λ_2 the system can present several distinct solutions. This qualitative change intervenes beyond a critical threshold, and it allows us to understand how a Bénard cell can be either left- or right-handed, as well as a host of other experimental facts surveyed in Chapter 1.

A striking difference between linear and nonlinear laws is whether the property of superposition holds or breaks down. In a linear system the ultimate effect of the combined action of two different causes is merely the superposition of the effects of each cause taken individually. But in a nonlinear system adding a small cause to one that is already present can induce dramatic effects that have no common measure with the amplitude of the cause. For instance, suppose that in Figure 23(b) the parameter λ changes first from 0 to a value just to the left of the limit point λ_2 of the lower branch (this is the "first cause"), and subsequently by a slight amount that brings it just to the right of this point (this is the "second cause"). With $\lambda = 0$ as a starting point, each of these two causes acting alone gives a very small "effect," in the form of a very slight change in the level of the variable X. But under their combined action the system will have to quit the lower branch of the response curve of Figure 23(b), as the latter is no longer defined beyond λ_2. The only issue left is then to perform a finite jump to the upper branch. This is clearly a very strong response, considering the smallness of the variation of λ induced by the second cause.

Note that without the maintenance of an appropriate distance from equilibrium, nonlinearity cannot by itself give rise to multiple solutions. At equilibrium detailed balance introduces a further condition that restricts and even uniquely fixes the value of the state variable. Moreover, by continuity this uniqueness should still extend to a domain of states near equilibrium. We can illustrate this property with a simple example involving the two coupled reactions

$$A + 2X \underset{k_2}{\overset{k_1}{\rightleftharpoons}} 3X$$

$$X \underset{k_4}{\overset{k_3}{\rightleftharpoons}} B$$

(2.15)

Here the concentration, x, of product X is taken to be the only state variable, it being understood that A and B are continuously supplied from or removed to the outside to maintain fixed concentrations, a and b. At equilibrium, detailed balance implies that

$k_1ax^2 = k_2x^3$ (equilibration of forward and backward steps in the first reaction)

$k_3x = k_4b$ (equilibration of second reaction)

These relations fix x_{eq} uniquely and in addition, impose a condition on the concentrations a and b of constituents A and B:

$$x_{eq} = \frac{k_4 b_{eq}}{k_3} = \frac{k_1 a_{eq}}{k_2}$$

$$(b/a)_{eq} = \frac{k_1 k_3}{k_2 k_4}$$

(2.16)

On the other hand, in a stationary state far from equilibrium it is sufficient to cancel the overall effect of the two forward reactions by that of the backward reactions, yielding

$$-k_2 x_s^3 + k_1 a x_s^2 - k_3 x_s + k_4 b = 0$$

(2.17)

This is a cubic equation for x_s that can have up to three solutions for certain values of a and b, as in Figure 2.3(b) and unlike what happens in the case of equilibrium. We can therefore say that nonequilibrium reveals the potentialities hidden in the nonlinearities, potentialities that remain dormant at or near equilibrium.

The possibility of having multiple solutions in a nonlinear system raises the question of *choice* between the different outcomes: clearly, a given Bénard cell cannot be both left- and right-handed. To understand this crucial aspect, certain supplementary notions that will be analyzed in the subsequent sections prove to be necessary. First, we identify some important nonlinearities encountered repeatedly in various fields.

Fluid mechanics   Fluid mechanics is full of nonlinear problems, essentially because the rate of change of a given property depends on, among other factors, the transport of that property by the natural flow of the fluid, whose velocity v is one of the variables of the problem. Thus the equation of conservation of momentum involves the manifestly nonlinear term $v \cdot \nabla v$. Moreover, in many problems the temperature T and the mass density ρ are also among the unknown variables. Additional nonlinearities arise through the coupling of these variables to the convection velocity, represented in the mass and internal energy balance equations respectively by the terms $\nabla(\rho v)$ and $v \cdot \nabla T$.

Chemistry and biology   In chemistry and biology nonlinearities may arise intrinsically, independent of spatial inhomogeneities. Most of the chemical reactions are the results of collisions. For instance, $X + Y \rightarrow D$ implies that the molecules of species X and Y must interact, and this gives rise to a rate proportional to $c_X c_Y$, i.e., to a nonlinear function. Of particular interest are the nonlinearities related to *regulation*. Here a substance X may activate or inhibit the rate of its own production or of the production of another constituent, which in turn feeds back on the first substance. In ordinary chemistry two characteristic examples are autocatalysis, which is illustrated by the first step of Eq. (2.15), and the thermal feedback discussed in Section 1.4 [see Eq. (1.5) and Figure 11]. In biology, regulation is intimately related to the peculiar structure and reactivity of the enzymes. In many cases it gives rise to nonpolynomial nonlinearities of the form

$$\left(\frac{dc}{dt}\right)_{reg} \sim \frac{ac^m}{1 + bc^n}$$

resulting in activation if $m \geqslant n$ and inhibition if $m < n$. Similar mathematical expressions are encountered in other contexts, for example, in electrical circuits containing active elements and in radiative processes in climate dynamics.

2.5 THE MANY FACETS OF THE SECOND LAW

We have seen that nonlinearity combined with nonequilibrium constraints allows for multiple solutions and hence for the diversification of the behaviors presented by a system. So far we have looked at these solutions after they were established and have analyzed some of their characteristics. Now it is time to raise the question of *how* these solutions of the evolution laws, Eqs. (2.12), are attained in the course of time.

For conservative systems the answer to this question is remarkably simple, at least in principle: Evolution is entirely dictated by the initial conditions (as in Figures 8 and 20), which fix the various constants of motion and hence the trajectories themselves through such relations as Eqs. (2.6). In practice, however, this works in a useful way only for relatively simple situations. We postpone the discussion of other cases until Chapter 3.

It is in the case of dissipative systems that the search for criteria of evolution leads to some deep and unexpected results. Indeed, the noninvariance of the equations for the state variables with respect to time reversal suggests that such systems may be characterized by an *irreversible* approach toward a final state, usually expected to correspond to a time-independent regime.

For isolated systems, in which no exchanges with the environment are allowed, this irreversible trend is expressed by the second law of thermodynamics. According to this law there exists a function of the state variables of the system that varies monotonically during the approach to the unique final state of thermodynamic equilibrium. Usually this *state function* is chosen to be the entropy, S, a quantity first introduced in the study of thermal machines and subsequently studied at a more microscopic level by Ludwig Boltzmann, Josiah W. Gibbs, and others. The statement of the second law is:

$$\frac{dS}{dt} \geq 0 \qquad \text{(isolated system)} \qquad (2.18)$$

It is usually interpreted as a tendency to increased disorder, since the microscopic theory associates entropy with the number of states that can be realized, given the conditions applied to the system. Indeed, we may argue that the more restricted the number of these states, the more ordered the system will be. For example, a solid near absolute zero has a low entropy, whereas a gas at ordinary temperature and pressure in which particles collide and move fast in all possible directions has a high entropy.

The above interpretation of entropy and the second law is fairly obvious for systems of weakly interacting particles, to which the arguments developed by Boltzmann referred. For instance, both experiment and common intuition suggest that from an initial situation in which only half the volume of a box is filled with interacting particles, Figure 24(a), there will evolve a situation in which matter is distributed more or less uniformly throughout the box, Figure 24(b). In such an evolution disorder is clearly increasing, since the later uniform state is globally less structured than the more differentiated initial state. And since the volume eventually occupied by the particles is twice as large as the initial volume, the number of microstates compatible with the same macroscopic configuration has also increased. Thus if we define entropy to be a measure of the number of accessible microstates, we realize that the evolution depicted in Figure 24 is nicely summarized by Eq. (2.18), which indicates an irreversible trend to maximum disorder.

For strongly interacting systems the above interpretation does not apply in a straightforward manner since, for one thing, we know that for such systems there exists the possibility of evolving to more ordered states through the mechanism of phase transitions.

Whatever the detailed microscopic definition of entropy for such systems might be, we expect that in the absence of external constraints the state of the system—

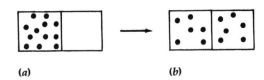

(a)   (b)

Figure 24 Spontaneous homogenization when a gas initially contained in the left half of a box (a) is allowed to occupy the whole volume (b). Both entropy and disorder increase during this evolution.

the thermodynamic equilibrium—will be dominated entirely by the interactions among molecules. It follows that in an isolated physical system in which gravitational interactions are not the main part of the dynamics, after an initial transient state described by relation (2.18) during which the system is dominated by irreversible processes, a final state of equilibrium is achieved in which the characteristic scale of spatial order will necessarily be microscopic, on the order of a few angstroms. A typical example of this kind of order is the crystal, in which the dimensions of the lattice constant are indeed of the order of the molecular diameter. Clearly a Bénard cell or a chemical clock could never arise spontaneously in such a state.

Let us now turn to nonisolated systems, which exchange energy or matter with the environment. The entropy variation will now be the sum of two terms. One, entropy flux, d_eS, is due to these exchanges; the other, entropy production, d_iS, is due to the phenomena going on within the system. Thus the entropy variation is:

$$\frac{dS}{dt} = \frac{d_iS}{dt} + \frac{d_eS}{dt} \qquad (2.19a)$$

For an isolated system $d_eS = 0$, and Eq. (2.19a) together with Eq. (2.18) reduces to $dS = d_iS \geqslant 0$, the usual statement of the second law. But even if the system is nonisolated, d_iS will describe those (irreversible) processes that would still go on even in the absence of the flux term d_eS. We thus require the following extended form of the second law:

$$\frac{d_iS}{dt} \geqslant 0 \qquad \text{(nonisolated system)} \qquad (2.19b)$$

As long as d_iS is strictly positive, irreversible processes will go on continuously within the system. Thus, $d_iS > 0$ is equivalent to the condition of dissipativity discussed in Section 2.2. If, on the other hand, d_iS reduces to zero, the process will be reversible and will merely join neighboring states of equilibrium through a slow variation of the flux term d_eS.

Among the most common irreversible processes contributing to d_iS are chemical reactions, heat conduction, diffusion, viscous dissipation, and relaxation phenomena in electrically or magnetically polarized systems. For each of these phenomena two factors can be defined: an appropriate internal flux, J_k, denoting essentially its rate, and a driving force, X_k, related to the maintenance of the nonequilibrium constraint. A most remarkable feature is that d_iS becomes a bilinear form of J_k and X_k, at least for the range of phenomena that are amenable to a local description similar to that offered by hydrodynamics or chemical kinetics:

$$\frac{d_iS}{dt} = \sum_k J_k X_k \qquad (2.20)$$

Table 2 summarizes the fluxes and forces associated with some commonly observed irreversible phenomena.

In general, the fluxes J_k are very complicated functions of the forces. A particularly simple situation arises when their relation is linear,

$$J_k = \sum_l L_{kl} X_l \qquad (2.21)$$

in which L_{kl} denote the set of *phenomenological coefficients*. This is what happens near equilibrium where, in addition, $L_{kl} = L_{lk}$, as shown by Lars Onsager on the basis of arguments related to the validity of detailed balance at equilibrium (see the suggested readings). Note, however, that certain states far from equilibrium can still be characterized by a linear dependence of the form of Eq. (2.21) that occurs either accidentally or because of the presence of special types of regulatory processes. In either case, by substituting the fluxes that appear in the evolution equations, Eqs. (2.12), in terms of the forces, and by realizing that the latter can in turn be expressed linearly in terms of the state variables themselves, we arrive in an alternative way at the conclusion formulated in Section 2.4: near equilibrium the nonlinearities that may be present in a system become ineffective.

Let us turn again to the general form, Eq. (2.19a). We have emphasized that the second law imposes $d_iS \geq 0$. In contrast to this restriction, in a nonisolated system ($d_eS \neq 0$) there exists no physical law that imposes the sign of d_eS. That is, d_eS can be positive or negative, depending on the system considered. It is therefore conceivable that d_eS can become sufficiently negative and exceed the

Table 2 Thermodynamic fluxes and forces in some commonly observed irreversible phenomena*

Phenomenon	Flux	Force	Vector character
Heat conduction	Heat flux, J_{th}	grad $(1/T)$	Vector
Diffusion	Mass flux of constituent i, $J_{d,i}$	$-[grad\,(\mu_i/T) - F_i]$	Vector
Viscous flow	Dissipative part of pressure tensor, P	$(1/T)$ grad v	Tensor (2d rank)
Chemical reaction	Rate of reaction ρ, w_ρ	Affinity of reaction divided by T: $\mathscr{A}_\rho/T$	Scalar

* T = temperature; μ_i = chemical potential of constituent i; F_i = external force per unit mass acting on i; v = hydrodynamic velocity. The affinity $\mathscr{A}_\rho$ is related to μ_i by $\mathscr{A}_\rho = -\sum_i \nu_{i\rho}\mu_i$, in which the stoichiometric coefficients $\nu_{i\rho}$ give the overall number of molecules produced ($\nu > 0$) or consumed ($\nu < 0$) in the reaction.

magnitude of d_iS, in which case certain stages of the evolution will be characterized by

$$dS/dt < 0 \qquad \text{(nonisolated system)} \qquad (2.22)$$

According to the traditional interpretation this would mean that because of the entropy flow, disorder has decreased in the course of the evolution. This would open the way to understanding the origin of the complex phenomena surveyed in Chapter 1 on a thermodynamic basis.

The situation is not that simple, however. As emphasized in the beginning of this section, the connection between entropy and order is clear in ideal (non-interacting) systems, or in more general systems at equilibrium, when the characteristic space scale of the ordered structures that may be formed is comparable to the scale of the intermolecular forces. On the other hand, the order associated with the emergence of dissipative structures under nonequilibrium conditions need not be related in a simple way to a decrease of entropy. We should thus regard Eq. (2.22) as simply a strong motivation to search for a sharper, less ambiguous characterization of complexity.

2.6 STABILITY

Let us express the content of the second law as applied to an isolated system from another point of view. On one side, it is asserted that whatever the initial preparation of a system, a unique state—the equilibrium state—will be attained in time. On the other side, it is argued that this trend will imply a monotonic increase of entropy in time or, alternatively, a monotonic increase of the excess $\Delta S = S - S_{eq}$ from negative values to zero, as shown in Figure 25.

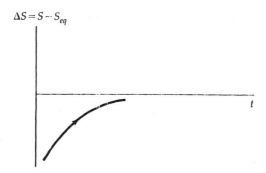

Figure 25 An alternative view of the second law of thermodynamics for an isolated system: the entropy excess ΔS is negative and increases monotonically in time, thus playing the role of a Lyapounov function. The equilibrium state is asymptotically stable and constitutes a global attractor.

Both statements are in fact equivalent, and in more mathematical language they amount to saying that equilibrium is an *asymptotically globally stable* state and that entropy is a *Lyapounov function* expressing this extremely strong form of stability. Let us define these terms more precisely, for they will play an important role in our subsequent analysis (also see Sections 1.3 and 1.4, where asymptotic stability was invoked in an intuitive way).

Consider a system—dissipative or conservative—evolving according to certain dynamical laws given, say, by Eqs. (2.12). By a mechanism that we need not specify, the system has arrived at a particular state X_S, in which it stays thereafter. In a conservative system X_S could be a state of mechanical equilibrium, whereas in a dissipative system it could represent the state of thermodynamic equilibrium or a stationary nonequilibrium state.

In actuality, a real-world system never stays in a single state as time varies. To begin with, most systems are in contact with a complex or even unpredictable environment that continuously communicates to them slight (or more rarely strong) quantities of matter, momentum, or energy. As a result, it becomes practically impossible to control any of the state variables with unlimited precision. Every experimenter encounters this restriction, and the time-honored terms "experimental error" and "confidence interval" acknowledge the interference of the environment with the intrinsic dynamics of the system being investigated. We express this ubiquitous fact by saying that the instantaneous state of our system is not X_S but, rather, a nearby state X related to X_S through

$$X(t) = X_S + x(t) \tag{2.23}$$

The quantity x is referred to as the *perturbation*.

There is a second, more fundamental view of the same problem. Most of the systems we are dealing with are composed of a large number of interacting entities. For such systems the state variables we have been referring to throughout this chapter must be understood in a statistical sense: they represent either averages of the instantaneous states over a long time interval, or perhaps the most probable among the values that the instantaneous state variables can reach. In actual fact, a recording of the instantaneous state will show continuous deviations from these reference values of the macroscopic state variables, deviations that the system generates spontaneously, independent of the environment. We call these intrinsic deviations *fluctuations*. Their effect can still be expressed through Eq. (2.23), provided it is understood that $x(t)$ is now part of the internal dynamics of the system considered.

Typically, a fluctuation originates in the form of a localized, small-scale event. For instance, in a small part of a macroscopic system certain particles can have a large thermal velocity compared to their neighbors. Shortly thereafter the density of particles in a different small subvolume can be less than the value in the surroundings, and so forth. We can easily understand that such events will essentially be random. Such terms as "background noise," frequently used to denote fluctuations, represent this property.

We have explained the reasons for which the state of a system will deviate continuously from the *reference state*, X_S. Now let us discuss the response of the system to those deviations. The following cases constitute a complete list of all possibilities.

Case 1 The state $X(t)$ remains in some vicinity of the reference state X_S for all times larger than a certain initial time t_0. More precisely, for any prescribed $\varepsilon > 0$, we can find an initial perturbation smaller than δ, where δ depends in general on ε and t_0, such that the deviation $x(t)$ remains smaller than ε for all times larger than t_0. We say then that the state X_S is *stable in the sense of Lyapounov*. Lyapounov stability is illustrated in Figure 8, in the diagram showing that a pendulum "remembers" the effect of a perturbation. Figure 26 provides three alternative views of this idea.

In many cases we are not concerned with the response of a given state to the perturbations, but with the response of a whole sequence of states defining a trajectory. We then introduce a suitable measure of the distance between the reference and perturbed trajectories and apply to it the above definition of stability. This leads to the notion of *orbital stability*.

Case 2 The state $X(t)$ tends back to X_S as time tends to ∞; in other words, the perturbation $x(t)$ decays to zero in time. We say then that X_S is *asymptotically stable*. Asymptotic stability is illustrated by the diagram in Figure 8 showing the response of a system that does not "remember" the effect of a perturbation. Further aspects of this are illustrated in Figure 27. The definition can be extended to *asymptotic orbital stability*. It is easy to see that asymptotic stability necessarily implies irreversibility and that for this reason it cannot apply to conservative systems. The latter can at best be stable in the sense of Lyapounov. On the other hand, dissipative systems are capable of eliminating the effect of perturbations that may act on them and thus can reestablish the reference state. This ensures the predictability and reproducibility of what is termed an *attractor*. For instance, the flow in Bénard cells, the oscillations in the BZ reagent, and the state of thermodynamic equilibrium in a isolated system are all attractors of appropriately defined dynamical systems. Undoubtedly, asymptotic stability is one of the most striking manifestations of the constructive role of irreversibility in nature.

Case 3 The state $X(t)$ does not remain in the vicinity of X_S. More precisely, in each neighborhood of X_S there exists an initial perturbation whose magnitude $x(t)$ cannot remain less than a prescribed but arbitrary value for all times larger than t_0. We say then that the reference state X_S is *unstable*. This situation is manifested through an initial stage of rapid (frequently exponential) growth of the perturbation. All these features can easily be extended to define *orbital instability*.

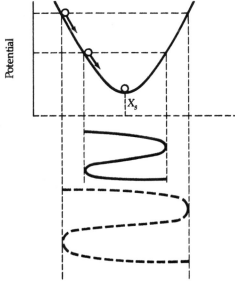

(a)

(b)

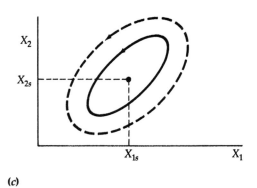

(c)

Figure 26 Three views of Lyapounov stability. (a) Perturbations around the reference state X_s remain bounded for all values of time. (b) Mechanical visualization of (a), in which the motion of a material point in a potential well is represented for two different initial conditions. (c) Phase space orbits corresponding to two different perturbations from the reference state (X_{1s}, X_{2s}). Figure (c) also illustrates the notion of orbital stability.

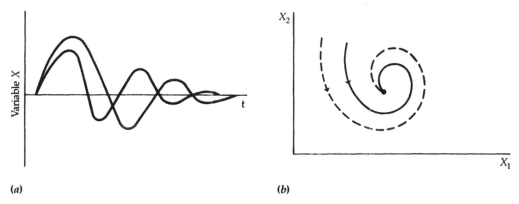

Figure 27 Two views of asymptotic stability. (a) Time evolution of state variable X. (b) Phase space trajectories converging to the attractor.

One manifestation of instability would be expressed by a diagram similar to that of Figure 27(b) but with the arrows reversed, indicating movement in the opposite direction. Another form of instability is illustrated in Figure 28.

Unstable reference states can occur in both conservative and dissipative systems. The state of equilibrium on the upward vertical in the pendulum and the state of rest in the Bénard problem beyond the threshold for convection are two examples. However, whereas no restriction applies to the dissipative case, in a conservative system the state must remain unstable upon time reversal. This is compatible with Figure 28(a) and (c) but would be in contradiction with a reversed-arrow version of Figure 27(b).

Case 4   The state $X(t)$ remains in some vicinity of the reference state X_S for values of initial perturbations less than a certain threshold, but moves away from X_S for perturbations beyond this threshold. We say then that X_S is *locally stable,* but *globally unstable.* If, on the other hand, stability prevails for any value of initial perturbation, we speak of *global stability,* the state X_S being in this case a *global attractor.* For instance, thermodynamic equilibrium in an isolated system is a global attractor, but the state of convection in the Bénard problem in which a particular cell is right-handed is globally unstable with respect to a finite perturbation producing a left-handed flow in the same cell. A mechanical representation of local stability and global instability is given in Figure 29. A mobile object with a small amount of kinetic energy starting to move (without friction) near the bottom of valley 1 will remain within it, but given a sufficiently large amount of kinetic energy it will overcome the "hilltop" and will follow a path that includes both valleys 1 and 2. In other words, the equilibrium positions in valley 1 and valley 2 are stable only toward perturbations that do not lead over the hill.

In the light of our discussion of stability, we can now understand better the conceptual advance achieved by the passage from isolated systems, dominated by

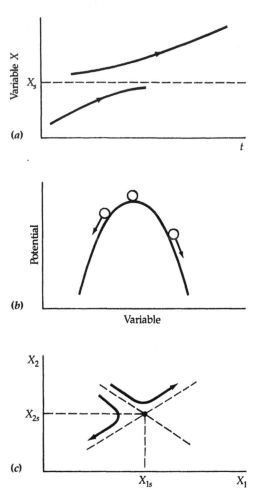

Figure 28 Three illustrations of the instability of a reference state. (a) Time evolution of a state variable X. (b) A mechanical illustration provided by the motion of a material point initially positioned on top of a relief. (c) Phase space representation. Notice that inverting the arrows in (a) and (c)—which is equivalent to performing a time reversal in the evolution laws—leaves the reference state unstable.

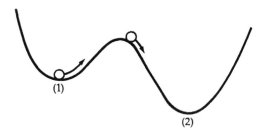

Figure 29 Global versus local stability. A mobile starting in valley (1) may remain within it or jump to valley (2) depending on whether its initial kinetic energy is small or exceeds a finite threshold.

relation (2.18), to open systems far from equilibrium, dominated by relations (2.19) and (2.20).

In an isolated system equilibrium is globally stable, and this is expressed by the existence of a function, the entropy—or more precisely, the excess entropy $\Delta S = S - S_{eq}$ (Figure 25) with the properties

$$\Delta S \leqslant 0, \qquad \frac{d}{dt}\Delta S \geqslant 0 \qquad \text{(isolated system)} \qquad (2.24)$$

We call ΔS a *Lyapounov function*. In the mathematical literature it is shown that Eq. (2.24) is indeed an expression of global stability.

In a nonisolated system not at equilibrium, because of the two terms $d_e S$ and $d_i S$ in the entropy balance, the second law ($d_i S/dt \geqslant 0$) no longer imposes the sign of entropy variation, nor in fact of any other state function. Thus, there is no universal Lyapounov function, and this raises the problem of the stability of states far from equilibrium. The possibility of loss of stability under certain conditions opens the way to transition phenomena that may lead to internal differentiation and complex behavior. One exception is worth noting, however, and that is an open system near equilibrium. Because of the linearity of the phenomenological laws and the symmetry of the coefficients in this range—Eq. (2.21)—the entropy production $d_i S/dt$ in Eq. (2.20) becomes a positive definite quadratic form of the constraints X_k. Moreover, it is possible to show that any deviation from the steady state leads to a time evolution tending to decrease the value of $d_i S/dt$. In other words, we have a situation similar to Eq. (2.24), in which the entropy production plays the role of a Lyapounov function that guarantees the global stability of steady states close to equilibrium.

It should be emphasized that despite the absence of a general law guaranteeing stability of states far from equilibrium, it is still possible to find useful criteria for making more limited statements—for example, specifying *sufficient conditions* for stability. One such criterion is provided by the condition of positivity of the excess entropy production evaluated around a nonequilibrium reference state, and is discussed in detail in the monograph by Glansdorff and Prigogine listed in the suggested readings.

2.7 BIFURCATION AND SYMMETRY BREAKING

Let us look once again at the Bénard problem from a new angle. We have already insisted on the role of nonlinearities and on the importance of the constraint driving the system far from equilibrium. We also know that the different transitions observed experimentally take place when this latter parameter is modified by the investigator. We shall now sort out certain general features that are present in a large number of other transition phenomena in physico-chemical systems.

Figure 30 shows the way the vertical component of the velocity of the flow at a certain point is influenced by the constraint or, in more general language, the way the state variable X (or $x = X - X_S$) of the system is affected by the control parameter λ. This kind of graph is known as a *bifurcation diagram*.

For small values of λ only one solution—the state of rest in the Bénard experiment—is accessible. It is the direct extrapolation of thermodynamic equilibrium and shares with it the important property of asymptotic stability, since in this range the system is capable of damping internal fluctuations or external disturbances. For this reason we call this unique solution range the *thermodynamic branch* of states. But beyond a critical value, denoted by λ_c in Figure 30, we find that the states on this branch become unstable: the effect of fluctuations or small external perturbations is no longer damped. The system acts like an amplifier, moves away from the reference state, and evolves to a new regime, the state of (stationary) convection in the case of the Bénard experiment. The two regimes coalesce at $\lambda = \lambda_c$, but are differentiated for $\lambda > \lambda_c$. This is the phenomenon of *bifurcation*. We can easily understand why this phenomenon should be associated with catastrophic changes and conflicts. Indeed, at the crucial moment of transition, when $\lambda = \lambda_c$, the system has to perform a critical choice. In the Bénard problem the choice is associated with the appearance of a right- or left-handed cell in a certain region of space. In Figure 30 branches b_1 and b_2 represent the two possible alternatives. Nothing in the description of the experimental setup permits the observer to assign beforehand the state that will be chosen; only chance will decide, through the dynamics of fluctuations. The system will in effect scan the territory and will make a few attempts, perhaps unsuccessful at first, to stabilize. Then a particular fluctuation will take over. By stabilizing it, the system becomes a historical object in the sense that its subsequent evolution depends on this critical choice.

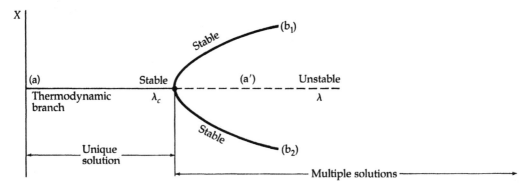

Figure 30 Bifurcation diagram showing how a state variable X is affected when the control parameter λ varies. A unique solution *(a)*, the thermodynamic branch, loses its stability at λ_c. At this value of the control parameter new branches of solutions (b_1, b_2), which are stable in the example shown, are generated.

In the above description we have succeeded in formulating, in abstract terms, the remarkable interplay between chance and constraint, between fluctuations and irreversibility, that underlay most of the phenomena surveyed in Chapter 1. Note the similarity between these ideas and the notion of mutation and selection familiar from biological evolution. As a matter of fact, we can say that fluctuations are the physical counterpart of mutants, whereas the search for stability is the equivalent of biological selection. Even the very structure of a bifurcation diagram is reminiscent of the phylogenetic trees employed in biology.

If instead of the Bénard flow we were interested in the Belousov-Zhabotinski reaction, the aggregation of *Dictyostelium discoideum*, or in practically any other transition phenomenon, and if we constructed bifurcation diagrams for these systems, we would find the same qualitative structure. The only differences would concern terminology, which is necessarily specific to each particular problem. Thus we find a deep unity among widely diverse systems, and this allows us to assert that the passage toward complexity is intimately related to the *bifurcation* of new branches of solutions following the *instability* of a reference state, caused by the nonlinearities and the constraints acting on an open system.

Figure 31 represents a mechanical illustration of the phenomenon. A ball moves in a valley, equivalent to branch *a* of Figure 30, which at a particular point λ_c becomes branched and leads to either of two new valleys, branches b_1 and b_2, separated by a hill. Although it is too early for analogies and extrapolations at this stage in our discussion, it is thought provoking to imagine for a moment that instead of the ball in Figure 31 we could have a dinosaur sitting there prior to the end of the Mesozoic era, or a group of our ancestors about to settle on either the ideographic or the symbolic mode of writing.

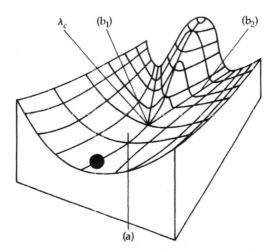

Figure 31 Mechanical illustration of the phenomenon of bifurcation.

We have emphasized that bifurcation is a source of innovation and diversification, since it endows a system with new solutions. But what are the characteristics of these solutions? We postpone the major part of the answer to this question until the next chapter. For the present we stress a particular point that arose in our discussion of experimental facts in Chapter 1, namely, that bifurcation generates solutions displaying *broken symmetries*.

It is a well-established fact that certain physical phenomena can be described in exactly the same way even if experiments are carried out on them under different observational conditions. For instance, a layer of liquid in a state prior to the threshold for Bénard convection has identical properties for all observers displaced to one another along a certain direction (translational invariance). Another familiar example, taken from mechanics, is that the acceleration of a body falling freely near the earth's surface will be vertical, not only for an observer fixed on the ground, but also for all observers moving with respect to him at arbitrary constant velocity (galilean invariance). These facts are important because they impose certain constraints that any reasonable mathematical model must respect. In particular, the laws describing the phenomena must display *symmetries* that allow for the *invariance* of the phenomena themselves. These symmetries must be preserved under all operations in space and time leading from a particular observer to any other observer for which the invariance holds true. Quite frequently these operations constitute a mathematical object known as a group, since the operations performed consecutively are equivalent to an operation that is itself compatible with invariance. Thus, the equations of fluid dynamics describing a liquid heated from below and contained between two plates that extend infinitely in the horizontal direction are invariant with respect to the group of translations along any horizontal plane. Similarly, the equations of chemical kinetics in an infinitely extended medium are invariant with respect to the groups of translations, rotations, and reflections.

Yet experiment shows that out of these extremely symmetric dynamical laws, we observe the generation of such states as a Bénard cell or a chemical wave, each of which is less symmetrical than the laws themselves. In each case a *symmetry-breaking* transition arose as a result of a bifurcation. What exactly did it achieve?

To begin with, symmetry breaking is the manifestation of an *intrinsic differentiation* between different parts of a system, or between the system and its environment. In this respect, it is one of the prerequisites of complex behavior and must have been at work during such events as the condensation of primordial matter to form galaxies, or the appearance of the first living cells. Second, once such a differentiation is ensured, further processes that would be impossible in an undifferentiated medium may be switched on. For instance, as illustrated by *Dictyostelium discoideum* in Chapter 1, the appearance of spatial inhomogeneities in a biological medium may enable undifferentiated cells in a population to recognize their environment and differentiate in specialized cells, thus allowing the genetic material to manifest its potentialities. In this respect, symmetry breaking appears to also be a prerequisite to *information*.

Still, there is a long way to go before a prerequisite becomes a sufficient condition. Indeed, symmetry-breaking transitions would be significant only if they could lead to asymptotically stable solutions. Above all, we should be able to *select* in one way or the other a particular asymmetric form of matter from among the multitude of the solutions that simultaneously become available beyond the bifurcation point. We discuss these matters further in the subsequent chapters.

2.8 ORDER AND CORRELATIONS

We have repeatedly referred to thermodynamic equilibrium as a state in which the characteristic space scales are microscopic, and to the transition phenomena arising far from equilibrium as being associated with states having correlations of macroscopic range. While the first statement appears natural in view of the short range of the intermolecular interactions, the second statement is disconcerting, to say the least. Why should changes in scale be associated with deviations from equilibrium? The apparent paradox is increased when we realize that nonequilibrium transitions encompassing, in a coherent fashion, a macroscopic region of space can even take place in ideal systems, in which the effect of intermolecular interactions can essentially be neglected!

In order to understand intuitively what is going on, it will be necessary to use the fluctuations as a probe to investigate the state of matter. Suppose first that we deal with an ideal system (e.g., a gas) at equilibrium. Let V and N be the total volume and the total number of particles, ΔV a small subvolume within V, and n the number of particles within ΔV. Since the system is uniform, the probability that any given particle is in volume ΔV is given by the ratio $\Delta V/V$, and the probability that n given particles are simultaneously in it is $(\Delta V/V)^n$. Similarly, the probability that a particle is *not* in ΔV is $(V - \Delta V)/V$, and the same probability for m particles is $[(V - \Delta V)/V]^m$. The probability P_n that the volume ΔV contains n particles is therefore

$$P_n = \text{(Number of ways of choosing } n \text{ out of } N \text{ particles)}$$
$$\times \left(\frac{\Delta V}{V}\right)^n\left(1 - \frac{\Delta V}{V}\right)^{N-n} \tag{2.25}$$

In the case of interest, ΔV is much smaller than V, and the number n is also small compared to N. The combinatorial factor in Eq. (2.25) may therefore be approximated by $N^n/n!$, where $n! = 1 \ldots n$.

Neglecting also n in the exponent $\cdot N - n$ of the second term, we write Eq. (2.25) as

$$P_n = \frac{1}{n!}\left(\frac{N\,\Delta V}{V}\right)^n\left(1 - \frac{\Delta V}{V}\right)^N$$

But $N \Delta V/V$ is simply the mean number of particles, $\bar{n}$, in the volume ΔV. Hence,

$$P_n = \frac{\bar{n}^n}{n!}\left(1 - \frac{\bar{n}}{N}\right)^N$$

Finally, realizing that N is very large so that the formula

$$\lim_{N \to \infty}\left(1 - \frac{x}{N}\right)^N = e^{-x}$$

can apply, we obtain

$$P_n = e^{-\bar{n}}\frac{\bar{n}^n}{n!} \tag{2.26}$$

This is the *poissonian distribution*, one of the important distributions of probability theory. Figure 32 shows that it is sharply peaked at a value close to $\bar{n}$. In order

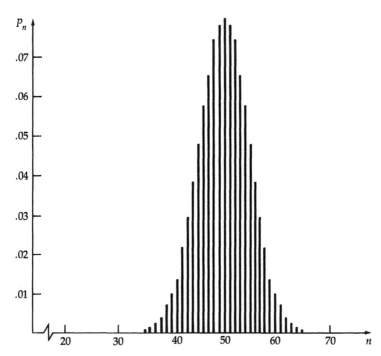

Figure 32 Poissonian probability distribution corresponding to $\bar{n} = 50$, plotted against n. Notice that the distribution is practically symmetrical around its sharp maximum at $n = \bar{n}$.

to characterize this sharpness, we introduce the *moments* $\langle n^k \rangle$ through

$$\langle n^k \rangle = \sum_{n=0}^{\infty} n^k P_n$$

In particular we find for $k = 0$ that

$$\sum_{n=0}^{\infty} P_n = 1,$$

in agreement with the meaning of probability (normalization condition), and for $k = 1$ that $\langle n \rangle = \bar{n}$, as expected. If P_n were infinitely sharp around $\bar{n}$, all subsequent moments would reduce to powers of $\bar{n}$. Conversely, their deviations from this power law will indicate the importance of the width of the distribution, and through it the importance of the fluctuations. From Eq. (2.26) we easily find that

$$\langle n^2 \rangle = \bar{n}^2 + \bar{n}$$

Hence, introducing the *variance* of P_n

$$\langle \delta n^2 \rangle = \langle n^2 \rangle - \langle n \rangle^2 \tag{2.27}$$

we find that for a poissonian distribution

$$\langle \delta n^2 \rangle_p = \bar{n} = \langle n \rangle \tag{2.28}$$

Moreover, from the derivation of Eq. (2.26) we see that the joint probability distribution for having n_1 and n_2 in two subvolumes ΔV_1, ΔV_2 (neighboring or not) is simply the product of the individual P_n's:

$$P_{n_1, n_2} = P_{n_1} P_{n_2} \tag{2.29}$$

Thus, taking the second moment $\langle n_1 n_2 \rangle$ of P_{n_1, n_2} we find

$$\langle n_1 n_2 \rangle = \langle n_1 \rangle \langle n_2 \rangle = \bar{n}_1 \bar{n}_2$$

Introducing the *covariance* of P_{n_1, n_2}

$$\langle \delta n_1, \delta n_2 \rangle = \langle n_1 n_2 \rangle - \langle n_1 \rangle \langle n_2 \rangle \tag{2.30}$$

we obtain

$$\langle \delta n_1 \, \delta n_2 \rangle_p = 0 \tag{2.31}$$

In short, we have arrived at two important quantities of statistical nature:

- The variance, which measures the importance of fluctuations relative to the mean
- The covariance, which measures the degree of correlation between different spatial regions in the system

In a poissonian regime such as that in the state of equilibrium of an ideal system, the vanishing of the covariance implies the complete absence of spatial correlations. Such a state will therefore be the prototype of disorder. This does not mean that fluctuations are large, however. Quite to the contrary, Eq. (2.28) shows that the quantity $\sqrt{\langle \delta n^2 \rangle_p / \bar{n}^2}$ —which can be regarded as a measure of the relative magnitude of the fluctuations—is equal to $1/\sqrt{\bar{n}}$, and hence is very small if the volume ΔV is large compared to the molecular size. For instance, if ΔV is one cubic centimeter (1 cm³), $1/\sqrt{\bar{n}}$ will be something like 10^{-8} percent for a gas under ambient conditions. However, for ΔV of one cubic micrometer (1 μm³) fluctuations will attain values of the order of a few hundredths of a percent.

Conversely, the emergence of a configuration of matter characterized by macroscopic space and time scales implies that the system generates and maintains reproducible relationships among its distant parts. This will be manifested by a spatial correlation function—the covariance $\langle \delta n(r_1)\, \delta n(r_2) \rangle$—which will display a finite amplitude and a macroscopic correlation length, in the sense that it will have appreciable values even for spatial distances $|r_2 - r_1|$ of macroscopic size. Both properties imply that the system has been able to overcome the "tyranny" of the poissonian distribution. How this is possible, even in situations in which the intermolecular interactions play only a minor role, will be discussed in Chapter 4.

In closing this chapter, let us stress that while we cannot yet attempt a clear-cut definition of complexity, we begin to perceive some of its essential ingredients: the emergence of bifurcations in far-from-equilibrium conditions, and in the presence of suitable nonlinearities; the generation of broken symmetries beyond bifurcation; and the formation and maintenance of correlations of macroscopic range. In the next two chapters these concepts will gradually be quantified, and as a result a more concrete picture of complexity will emerge.

Dynamical Systems and Complexity

Complexity is an inseparable part of the world of dynamical systems. Henri Poincaré, the great mathematical genius of the turn of the century, invented the modern theory of dynamical systems and set as an objective the exploration of the types of behavior that can be expected from systems described by coupled nonlinear equations. In this chapter we present some aspects of this theory that are relevant for the problem of complexity, with special emphasis on the minimal ingredients beyond which complex behavior can arise and on the principal characteristics of the solutions of the underlying equations.

3.1 THE GEOMETRY OF PHASE SPACE

In the first two sections of Chapter 2, we introduced the two basic families of dynamical systems encountered in nature, conservative and dissipative systems. We also explained how their time evolution can be visualized through the motion of a representative point in phase space. In this section we develop this idea further by examining the types of possible phase space motions. By virtue of the one-to-one correspondence between such motions and the behavior of the original system, we will achieve a preliminary qualitative classification of the phenomena that can be expected from a dynamical system.

We begin with dynamical systems involving a finite number of variables. This rules out, for now, spatially distributed systems described by variables that depend continuously on space. Such cases will be considered in the last two sections of this chapter. The evolution equations, Eqs. (2.12), take the form

$$\frac{dX_i}{dt} = F_i(\{X_j\}, \lambda) \qquad i = 1, \dots, n \tag{3.1}$$

in which it is now understood that there is no space dependence involved in the operator F_i. Moreover, we restrict ourselves to autonomous systems, for which F_i does not depend explicitly on time, a consequence being that the trajectories in phase space are invariant. Note that in a hamiltonian system n must be even and F_i must reduce to the characteristic structure imposed by Eq. (2.5b).

A first kind of phase space trajectory compatible with Eq. (3.1) is given by

$$\frac{dX_1}{dt} = \cdots = \frac{dX_n}{dt} = 0 \tag{3.2}$$

It includes as particular cases the states of mechanical equilibrium encountered in conservative systems and the steady states encountered in dissipative systems. In phase space such trajectories are quite degenerate, since they are given by the solutions of the n algebraic equations for n unknowns, $F_i = 0$. They are thus represented by points, such as P in Figure 33, that will be referred to as *fixed points*.

If Eqs. (3.2) are not satisfied, the representative point will not be fixed but will move along a phase space trajectory defining a curve, (a in Figure 33). The line element along this trajectory for a displacement corresponding to $(dX_1, \dots, dX_n)$ along the individual axes is given by

$$ds = \sqrt{\sum_k dX_k^2} = \sqrt{\sum_k F_k^2} \; dt.$$

Thus, the projections of the tangent of the curve along the axes are given by

$$dX_\alpha/ds = F_\alpha/\sqrt{\sum_k F_k^2}$$

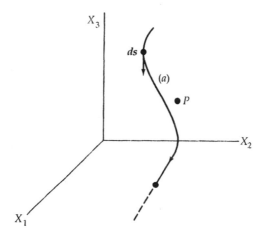

Figure 33 Phase space description of a dynamical system. P, fixed point; (a), phase space trajectory.

and are well defined everywhere. We call the points belonging to such curves *regular points*. In contrast, the tangent on the fixed points is ill-defined because of the simultaneous vanishing of all F_i's. We may therefore also refer to fixed points as the *singular points* of the flow generated by Eqs. (3.1). The set of fixed points and phase space trajectories will constitute the *phase portrait* of a dynamical system.

One property that plays a decisive role in the structure of the phase portrait relates to the uniqueness theorem of the solutions of ordinary differential equations. This important result, which goes back to the work of the French mathematician Augustin Cauchy, asserts that under quite mild conditions on the functions F_i the solution corresponding to an initial condition not on a fixed point exists and is unique for all times in a certain interval $(0, T_m)$, whose upper bound T_m depends on the specific structure of the functions F_i. In the phase space representation, the theorem automatically rules out the intersection of two trajectories in any regular point.

A second property of great importance is the existence and structure of *invariant sets* of the flow. By this we mean objects embedded in the phase space that are bounded and are mapped onto themselves during the evolution generated by Eqs. (3.1). An obvious example of an invariant set is the ensemble of fixed points. Another is a closed curve in phase space, such as curves a and a' of Figure 20 representing the pendulum. Once on such a curve, the system goes repeatedly through exactly the same states; in other words, a closed phase space trajectory represents a periodic motion.

Let us illustrate the power of the above two properties on dynamical systems evolving in a two-dimensional phase space. Suppose that in the part of the phase space considered there are only two invariant sets, a unique fixed point and a

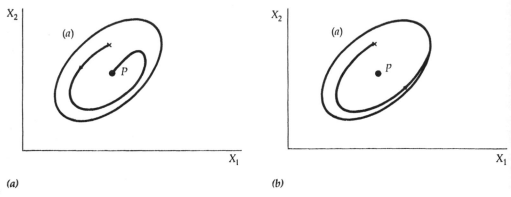

(a)   (b)

Figure 34 Role of invariant sets and nonintersection of trajectories in the structure of phase portrait. It is assumed that a closed trajectory *a* and a fixed point *P* are the only invariant sets available. A phase space trajectory emanating from a point inside curve *a* is bound to follow one of the two motions shown in (a) or (b).

closed curve surrounding this point—point *P* and curve *a* in Figure 34 (notice that this limits us automatically to dissipative systems). By the uniqueness theorem, a trajectory starting on an initial condition inside curve *a* will remain inside forever, otherwise it would have to intersect curve *a*. On the other hand, self-intersections are also ruled out by the same theorem. The only possibilities are then:

- To move on a curve tending to the fixed point *P*, Figure 34(a)
- To move on a curve tending to the closed curve *a*, Figure 34(b)

We see that the impossibility of self-intersection of the trajectories and the existence of invariant sets of a certain form determine, to a large extent, the structure of the phase space portrait in two-dimensional phase spaces, and through it the type of behavior that may arise. In three or more dimensions, however, the constraints imposed by these properties are much less severe, since the trajectories have many more possibilities to avoid each other by "gliding" within the "gaps" left between invariant sets. As we will see, new dynamical behaviors are possible because of this additional flexibility, not the least of which is the emergence of chaotic, turbulentlike evolutions.

3.2 MEASURES IN PHASE SPACE

According to the preceding discussion, the evolution of a dynamical system can be viewed as a transformation mapping the phase space onto itself. In the course

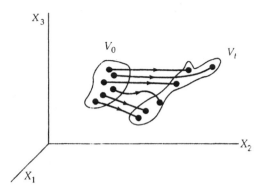

Figure 35 Change of an initial volume V_0 in phase space as representative points inside this volume evolve under the effect of the dynamics.

of this transformation, trajectories emanating from a certain part of phase space may visit a nearby region, reach other remote parts of this space, and perhaps remain confined there forever; or they may come back to the part of phase space from which they started. As a rule, different parts of the space will be visited with unequal frequency during these processes. The question now is how to characterize this complicated motion in a manner that is as compact and informative as possible.

One possibility is to attribute a "weight," a quantitative value, to each part of phase space and see how it is affected by the evolution. A natural weight is the volume of the part in question. To see how the evolution changes it, we should consider all trajectories starting with an initial condition inside this part, follow them during a time interval t, and reconstitute the part at this shifted time t as a set containing the representative points of the trajectories. This is done in Figure 35, and in this way we obtain a volume V_t at time t, whose initial value we denote by V_0.

Figure 36 describes three characteristic cases giving rise to different types of change of V_t compared to V_0. Case (a), a uniform circular motion, is the simplest. The equations of motion derive from the hamiltonian $H = (X_1^2/2) + (X_2^2/2)$. Not only is the phase space volume (shaded area) conserved, $|V_0| = |V_t|$, but because of the isochrony of the movement on each trajectory, it is subject to no deformation whatsoever.

Case (b) in Figure 36 describes a hyperbolic motion whose trajectories have the axes X_1 and X_2 as asymptotes. The corresponding hamiltonian is $H = X_1 X_2$. The projection of the movement along the X_1 axis gives rise to an exponential growth, whereas along the X_2 axis it gives rise to an exponential decay with the same absolute value of the characteristic exponent. Because of this the phase plane

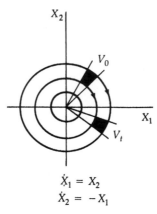

$$\dot{X}_1 = X_2$$
$$\dot{X}_2 = -X_1$$

Conservative *(a)*

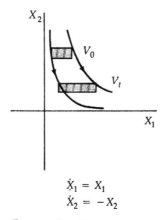

$$\dot{X}_1 = X_1$$
$$\dot{X}_2 = -X_2$$

Conservative *(b)*

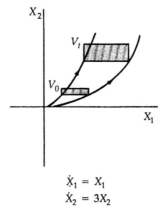

$$\dot{X}_1 = X_1$$
$$\dot{X}_2 = 3X_2$$

Nonconservative *(c)*

Figure 36 Types of change experienced by an initial volume V_0 for three representative examples of dynamical systems.

volume is again conserved, $|V_t| = |V_0|$, but contrary to the previous case, it is highly deformed. For instance, a rectangle transported by the motion will expand along its side parallel to X_1 and contract along the other side. This feature is characteristic of the class of *unstable motions* that will be discussed later in this chapter as well as in Chapter 5.

Case (c) also describes an unstable motion. However, since there is expansion along both X_1 and X_2, the phase space volume is not only deformed, it also increases systematically. That is, $|V_t| > |V_0|$. Clearly, such an evolution cannot go on forever, as it will eventually lead to an explosion. If, on the other hand, the signs of the righthand sides of the equations of evolution were negative rather than positive, we would obtain the same trajectories as in Fig. 36(c) but the arrows indicating the direction followed along the trajectories would be reversed. In this stable motion the phase space volume would again be deformed, but it would decrease systematically; that is, $|V_t| < |V_0|$. We call this latter system a *dissipative system*.

Whereas no general quantitative statement concerning the change of V_t can be made in a dissipative system, for a conservative system it turns out that the result $|V_t| = |V_0|$ found in the two examples of Figure 36(a)(b) is not an accident. A very important theorem, due to the French mathematician Joseph Liouville, establishes that the phase space volume remains invariant in *any* conservative system.

Before we discuss the implications of this result, let us look at the deformation of phase space volume from a different point of view. In principle the solution of the evolution Eqs. (3.1) constitutes a well-posed problem, in the sense that a complete specification of the state $(X_1, \dots, X_n)$ at any one time allows prediction of the state at all other times. But in many cases such a complete specification may be operationally meaningless. For example, in a hamiltonian system composed of particles whose number is of the order of Avogadro's number, there is no physical experiment giving access to the coordinates and velocities of all particles. More to the point, when the motion in phase space becomes very complex, as in a chaotic regime (also see Sections 3.7 to 3.9), it is no longer meaningful to argue in terms of individual trajectories. New modes of approach are needed, and among them the most widespread is certainly a description of the system in terms of probability concepts. For this purpose we consider not the rather special case of a single system, but instead focus attention on an ensemble of a very large number of identical systems, all subject to exactly the same constraints. The members of this ensemble, known as the *Gibbs ensemble* after the American scientist who introduced this concept, will in general be in different states. They can therefore be regarded as emanating from an initial ensemble of systems whose representative phase points were contained in a certain phase space volume V_0, as in Figure 35. Nothing in the above definition limits the number of systems in the ensemble, so it is convenient to consider the limit in which each region, however small, contains a large number of systems, and to ask for the fraction of the cases in the ensemble when some of the variables $\{X_i\}$ assume values in the ranges $\{X_i, X_i + dX_i\}$. In the continuous limit this will be expressed in terms of a *probability density*, $\rho(X_1, \ldots, X_n, t)$ such that

$$\rho \, dX_1 \cdots dX_n \tag{3.3}$$

represents the probability of finding at time t a member of the ensemble in the volume element $dX_1 \cdots dX_n$ of phase space surrounding the state $\{X_1, \cdots, X_n\}$. We require that ρ be nonnegative and normalized,

$$\int \rho \, dX_1 \cdots dX_n = 1 \tag{3.4}$$

In order to predict the probability of occurrence in the ensemble of particular values of certain variables, we must set up an equation of evolution for ρ. To this end we argue in terms very similar to fluid mechanics, since ρ can be regarded as the density of a fluid whose streamlines are the trajectories described by Eqs. (3.1).

Consider a fixed volume element of phase space located between X_1 and $X_1 + dX_1$, X_2 and $X_2 + dX_2$ as in Figure 37. As the X_i's vary in accordance with Eq. (3.1), the number of systems located in the volume $dX_1 \, dX_2 \cdots$ will change in time because systems will enter and leave this volume through its boundaries. For instance, the systems entering through the face $X_1 = const$ during the time interval dt are contained in the volume $\dot{X}_1 \, dt \, dX_2 \cdots dX_n = F_1 \, dt \, dX_2 \cdots dX_n$, and their number is thus given by

$$\rho(X_1, \ldots, X_n, t)F_1 \, dt \, dX_2 \cdots dX_n$$

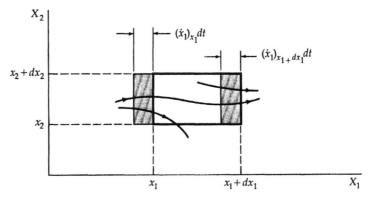

Figure 37 Illustration of the balance of the number of systems of a Gibbs ensemble (shaded area) entering into a fixed phase space volume element from the left and subsequently leaving this volume to the right.

The number of systems leaving through the face $X_1 + dX_1 = const$ during the same time interval is given by essentially the same expression, except that ρ and F_1 must be evaluated at $X_1 + dX_1$ instead of X_1:

$$\rho(X_1 + dX_1, \ldots, X_n, t)F_1(X_1 + dX_1, \ldots, X_n)\, dt\, dX_2 \cdots dX_n$$

$$= \rho(X_1, \ldots, X_n, t)F_1\, dt\, dX_2 \cdots dX_n + \frac{\partial \rho}{\partial X_1} F_1\, dt\, dX_1 \cdots dX_n$$

$$+ \rho\frac{\partial F_1}{\partial X_1}\, dt\, dX_1 \cdots dX_n$$

where the last part of the equality follows from a series expansion of $\rho(X_1 + dX_1)$ and $F_1(X_1 + dX_1)$ around X_1 in which only the first nontrivial terms are retained.

The same reasoning can now be repeated for the other boundaries of the volume, thus leading to the equation

(Change of ρ during dt) $dX_1 \cdots dX_n$ = Number of systems entering the volume

$-$ number of systems leaving the volume

or, dividing through by $dt\, dX_1 \cdots dX_n$ and using the above two expressions, for all faces of the volume,

$$\frac{\partial \rho}{\partial t} = -\sum_{i=1}^{n} \left(\frac{\partial \rho}{\partial X_i} F_i + \rho \frac{\partial F_i}{\partial X_i} \right) \tag{3.5}$$

Observe now that $\sum_i \partial F_i/\partial X_i$ can be viewed as the divergence of a vector $\mathbf{F}$ whose components are the rates of change $F_1 \cdots F_n$ of the individual variables X_i, Eq. (3.1). Moreover, if we follow the "fluid" of trajectories as time changes, we will reconstitute the "hydrodynamic" derivative $d\rho/dt$, in which both the explicit time dependence of ρ as well as its implicit time dependence through the variables $X_1(t) \cdots X_n(t)$ will be taken into account:

$$\frac{d\rho}{dt} = \frac{\partial \rho}{\partial t} + \sum_i \frac{\partial \rho}{\partial X_i} \dot{X}_i = \frac{\partial \rho}{\partial t} + \sum_i \frac{\partial \rho}{\partial X_i} F_i$$

We thus obtain from Eq. (3.5) the important result

$$\frac{d\rho}{dt} = \frac{\partial \rho}{\partial t} + \sum_{i=1}^{n} \frac{\partial \rho}{\partial X_i} F_i = -\rho \, div \, \mathbf{F} \tag{3.6}$$

If we evaluate div $\mathbf{F}$ for a hamiltonian system, we see from Eq. (2.5b) that

$$div \, \mathbf{F} = \sum_{i=1}^{n} \left(\frac{\partial}{\partial q_i} \dot{q} + \frac{\partial}{\partial p_i} \dot{p}_i \right) = \sum_{i=1}^{n} \left(\frac{\partial}{\partial q_i} \frac{\partial H}{\partial p_i} - \frac{\partial}{\partial p_i} \frac{\partial H}{\partial q_i} \right) = 0$$

Equation (3.6) implies then that the probability density ρ is conserved along the phase space motion. This is precisely Liouville's theorem expressed in a different language, and leads through Eq. (3.6) to the *Liouville equation*:

$$\frac{d\rho}{dt} = \frac{\partial \rho}{\partial t} + \sum_{i=1}^{n} \frac{\partial \rho}{\partial X_i} F_i = 0 \qquad \text{(conservative system)} \qquad (3.7)$$

Actually, Eq. (3.7) may remain valid for systems that are more general than hamiltonian systems. It suffices to have a flow in phase space in which the phase density behlves like the density of an incompressible fluid, div $\mathbf{F} = 0$. This condition can in fact be used to define the class of conservative systems. Similarly, Eq. (3.6) with div $\mathbf{F} \neq 0$ will be a necessary (though not always sufficient) condition that must be satisfied by the class of dissipative systems. For this latter class we require, in addition, that there be on the average a global contraction of volume in phase space, as the system moves toward its attractor.

The Liouville equation for conservative systems, Eq. (3.7), plays a very important role in statistical physics as it is the starting point for both the ergodic theory and the kinetic theory of irreversible processes. We discuss these questions more fully in Chapter 5.

For dissipative systems the use of statistical concepts, and in particular Eq. (3.6), turns out to be necessary in the case of chaotic behavior, in which a deterministic motion presents a markedly random character. These problems will be taken up in Section 3.10. In the next few sections we discuss conservative and dissipative systems whose dynamics involves a phase space of low dimensionality. Despite their apparent simplicity, it will turn out that some of these systems can present highly nontrivial modes of behavior, reminiscent of the phenomena surveyed in Chapters 1 and 2.

3.3 INTEGRABLE CONSERVATIVE SYSTEMS

Consider a hamiltonian system described by Eqs. (2.5) or (3.7). We set $n = 2N$, N being the number of inde endent positions or momenta of the problem, also referred to as the number of *degrees of freedom*. As is well known from calculus, the integration of a system of $2N$ ordinary differential equations of first order amounts to finding $2N$ independent first integrals or constants of motion, one of which provides a relation between the system's variables. Hamiltonian systems constitute, however, a particular class in which, under quite general conditions, it suffices to know only N first integrals. As Liouville demonstrated, if the latter are sufficiently regular in the mathematical sense of the term, the system can be integrated by simple quadratures.

Liouville's theorem tells more. Let I_i, $i = 1, \ldots, N$, be a suitable combination of the N constants of motion. Then, the hamiltonian flow can be reduced to a set of equations of the form

$$\frac{dI_i}{dt} = 0$$

$$\frac{d\varphi_i}{dt} = \omega_i(\{I_j\}) \qquad i = 1, \ldots, N$$

(3.8)

These equations have also a hamiltonian structure, by virtue of which the ω_i's are linked to the new (φ-independent) hamiltonian H through $\omega_i = \partial H / \partial I_i$. We call I_i and φ_i action and angle variables, respectively.

Equations (3.8) define the particular class of *integrable systems*. According to Liouville's theorem such systems can be characterized by either of the following two properties: Possess N sufficiently regular first integrals, or be amenable to Eqs. (3.8) by means of an appropriate transformation, known as canonical transformation, that preserves the hamiltonian structure.

As a corollary, any hamiltonian system with one degree of freedom ($n = 2$, $N = 1$) is integrable since it possesses one regular constant of motion, the total energy $H = E = const$. The pendulum—Eq. (2.6) and Figure 20—or the harmonic oscillator are therefore integrable systems. A hamiltonian system with two degrees of freedom is integrable if there exists a sufficiently regular first integral independent of the hamiltonian H. For three degrees of freedom three first integrals are needed. In the spinning top, which is one of the favorite three-dimensional problems treated in introductory mechanics, the existence of these integrals follows from conservation of energy and angular momentum. The spinning top is thus an integrable system. So too is the system in a two-body problem in the presence of central interaction forces, whose importance stems from its relation to the motion of celestial bodies.

More generally, all systems that can be separated into uncoupled systems of one degree of freedom are integrable. Systems of linear equations of motion are an obvious example, since a "normal mode" transformation diagonalizing the matrix of the coefficients reduces them to uncoupled, one degree of freedom equations. This is in fact at the basis of the extensive literature on small vibrations around a position of equilibrium, of which solid state physics in the harmonic approximation is a particularly important illustration. Certain nonlinear systems can also be integrable. The most interesting examples are spatially distributed systems like the Toda lattice or the Korteweg-de Vries equation.

Let us now study the kinds of behavior expected to arise in an integrable system. In the representation in which the evolution is described in terms of the constants of motion (actions) and the corresponding angles, the answer to this question is very simple. For one degree of freedom we obtain

$$I_1 = I_{10} = \text{const}$$

$$\varphi_1 = \omega_1 t + \varphi_{10}$$

(3.9)

If we look at (I_1, φ_1) as polar coordinates, in the corresponding orthogonal co-ordinates P, Q the problem would be isomorphic to the harmonic oscillator. The phase space is a plane, and for each value of I_1 the motion is uniform and takes place along the circumference of a circle, as in Figure 38.

For many degrees of freedom the motion is just the superposition of N independent motions, each one described by an equation of the type of Eq. (3.9). The phase space is $2N$-dimensional, but in view of $I_i = const$, the motion is confined on a set of N dimensions whose projection in the phase planes of the individual degrees of freedom is a circle. Such an object, called a torus, is visualized in Figure 39 for the case of $N = 2$. Notice that, in view of the conservation of the total energy, the phase space trajectories of a system with two degrees of freedom are confined in a three-dimensional subspace. The torus of Figure 39 is thus embedded in a three-dimensional space.

The time dependence of the variables P_i, Q_i follows immediately. For $N = 1$ it is periodic with period equal to $2\pi/\omega_1$, whereas for $N > 1$ it is the superposition of periodic functions of different periods. If these periods are commensurate, that is, if there exists a set of nonvanishing integers $k_1, k_2, \ldots$ such that

$$k_1\omega_1 + k_2\omega_2 + \cdots = 0$$

then the motion is clearly *periodic* and is represented by a closed curve. But if there exist no nonvanishing integers for which this equation can be satisfied, then

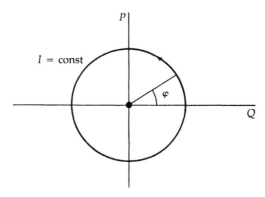

Figure 38 Phase space trajectory of a conservative system with one degree of freedom. Such a system is automatically integrable.

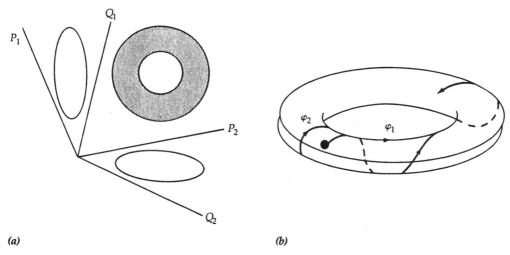

P_1
Q_1
P_2
Q_2

φ_2
φ_1

(a)   *(b)*

Figure 39 Visualization of phase space trajectories of an integrable dynamical system with two degrees of freedom. The motion takes place on the surface of a two-dimensional torus whose size depends on a combination of the action variables. The conservation of total energy reduces the number of independent variables from four to three. The torus in (b) is therefore embedded in a three-dimensional space.

the motion is *quasi-periodic* and is represented by a helix that moves on a torus without ever closing to itself. For $N = 2$ this would happen, for instance, if $\omega_1 = 1$ and ω_2 was an irrational, e.g., $\omega_2 = \sqrt{2}$.

A quasi-periodic motion can look quite intricate. Because its representative trajectory never closes and never intersects itself by virtue of the uniqueness theorem discussed in Section 3.1, it gradually "fills" the entire torus in the sense that it is eventually bound to pass through a prescribed neighborhood of any point of the toroidal surface. Such motions are called "everywhere dense."

As stressed repeatedly in Chapter 2, and especially in Section 2.6, a conservative system cannot enjoy asymptotic stability. Thus, neither the periodic nor the quasi-periodic motions described above can be used as models for attractors. It is easy to understand why: By the first relation, Eq. (3.9), there exists a continuum of admissible values for the constants of motion I_i. Thus instead of Figures 38 and 39(b) we have in fact a phase portrait involving a continuum of closed orbits or of toroidal surfaces nested inside each other. Inevitably, a small perturbation removing the system from a given orbit will throw it onto another invariant surface on which the system will remain trapped until another perturbation throws it onto a still different invariant surface, and so on. At best, therefore, under the effect of perturbations we expect simple or orbital stability and a jittery motion, rather than a sharply reproducible behavior.

Motions in which a continuum of closed invariant sets surrounds an equilibrium point are called *elliptic*, and the equilibrium point itself is referred to as the *center*.

According to the discussion of Section 3.2 [see especially Figure 36(a)], we expect that these motions will give rise to rather moderate deformations of the phase space volume (which of course will be conserved by Liouville's theorem).

So far we have described the behavior of integrable systems in the action-angle variable representation. These variables are related to the original ones by a transformation that in some exceptional cases is quite simple. For instance, in the harmonic oscillator problem

$$H = \tfrac{1}{2}m\dot{x}^2 + k\frac{x^2}{2}$$

I and φ are implicitly given through

$$x = \left(\frac{2I}{\sqrt{mk}}\right)^{1/2}\sin\varphi,$$

$$\dot{x} = \left(2I\sqrt{\frac{k}{m^3}}\right)^{1/2}\cos\varphi$$

and the hamiltonian reduces to

$$H = \sqrt{\frac{k}{m}}\,I = \omega I$$

ω being the angular frequency of the oscillator. In general, however, the transformation is highly complex. We expect therefore that in the original variables the trajectories will be highly deformed versions of the pictures shown in Figures 38 and 39(b). Moreover, certain types of motion occurring in the original phase space will become very artificial in the action-angle representation for the simple reason that the angle variable will cease to be an angle satisfying the condition $0 \leqslant \varphi \leqslant 2\pi$. This is the case of unstable motions, like the hyperbolic motion depicted in Figure 36(b), in which some of the variables become unbounded and the periods of certain trajectories tend to infinity. In the pendulum this occurs near the point of unstable equilibrium, $\theta = \pm\pi$ (see Figure 20). The region of bounded oscillations (vibrations) is separated from the region of unbounded motion (rotation) by the particular type of trajectory known as a *separatrix* (Figure 40), joining $\theta = \pi$ and $\theta = -\pi$. Since the rate of change of a variable goes to zero near a fixed point, the time needed to move from any point of the separatrix to the fixed point is infinity. The peculiar configuration of the trajectories near a separatrix, which looks like the level surfaces near a mountain pass, is at the origin of the term *saddle point* used to characterize unstable equilibria in conservative systems. As suggested in Figure 36, motion near a saddle point gives

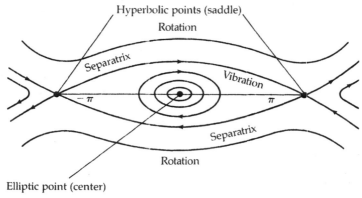

Figure 40 Phase portrait of the pendulum. Around the elliptic fixed point, which describes the unique position of stable equilibrium, a continuum of periodic vibrational motions whose amplitude and period depend on the initial conditions can take place. The hyperbolic fixed points at $\pm \pi$ describe the unstable equilibrium. They are joined by a closed loop of two separatrix segments. An initial condition outside this loop gives rise to a free rotational motion of the pendulum.

rise to large deformations of the phase space volume, while conserving its magnitude.

In short, aside perhaps from the separatrices and the saddle points—which bear in a hidden form a major mechanism of complex behavior—one can assert that the world of integrable systems is a simple and uneventful world. That is not so surprising, since in an integrable system we effectively get rid of the interactions and deal with independent entities. What is remarkable is that, despite their exceptional character, integrable systems dominated science until the 1950s, and they still constitute the main subject of most mechanics textbooks. Their great historical role and their indisputable pedagogical value are certainly partial explanations of this paradoxical situation. But for our purposes it is necessary to emphasize that the world of integrable systems cannot be the birthplace of complexity. We must look elsewhere, and that search is the main purpose of the rest of this chapter.

3.4 BIFURCATION IN A SIMPLE DISSIPATIVE SYSTEM: THE SEARCH FOR ARCHITYPES OF COMPLEXITY

We first turn our attention to dissipative systems. We choose a simple abstract mathematical model that will help us realize some of the mechanisms at work in a qualitative change of the behavior. Later we will relate it to real-world physico-chemical systems.

Consider a single variable x evolving according to

$$\frac{dx}{dt} = f(x, \lambda) = -x^3 + \lambda x \tag{3.10}$$

This simple system is controlled by a single parameter λ, on which the rate function f depends in a perfectly regular fashion. Let us first determine the fixed points (steady states):

$$-x_s^3 + \lambda x_s = 0$$

One solution of this simple algebraic equation that exists under any conditions is the trivial one,

$$x_0 = 0 \tag{3.11a}$$

Nontrivial solutions may exist, however. They satisfy the equation obtained by eliminating x_s from the above cubic:

$$-x_s^2 + \lambda = 0$$

If λ is negative this equation is meaningless, since it gives a complex-valued solution that cannot describe a physical situation. But if λ is positive the equation admits the pair of solutions

$$x_\pm = \pm\sqrt{\lambda} \tag{3.11b}$$

which coalesce with x_0 for $\lambda = 0$ but branch from it for $\lambda > 0$. This is the phenomenon of *bifurcation* discussed at length in Section 2.7 and illustrated in Figure 30; it is also known as *pitchfork bifurcation*. In Figure 41(a) a similar diagram is adapted to the structure of our specific problem. Solid and broken lines have been used to denote, respectively, asymptotically stable and unstable solutions. For the simple model considered, the stability problem is trivial, as Eq. (3.10) can be solved exactly by standard methods. We find that x_0 is globally asymptotically stable for $\lambda < 0$ and unstable for $\lambda > 0$, and that x_+, x_- are asymptotically (but not globally) stable. In other words, the branches $x_\pm$ bifurcate in the direction in which the reference state $x_0 = 0$ loses its stability, and are themselves stable. We call this *supercritical bifurcation*. Actually, the relation between supercritical branching and stability is not an accident. A general result of bifurcation theory, discussed in some detail in Appendix 2, establishes that under certain conditions on the rate function f frequently satisfied in practical applications, supercritical branches are stable and subcritical branches are unstable.

But the most striking result of our simple model is certainly illustrated by Eq. (3.11b). Remember that originally we had a simple cubic equation depending smoothly on the parameter λ. We now see that this very smooth dynamics gave rise to a *singularity*. Indeed, in the vicinity of the bifurcation point $\lambda = 0$ the

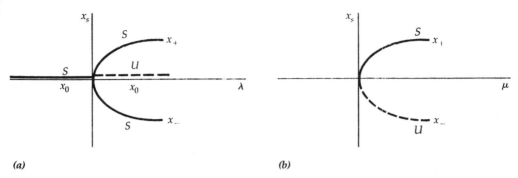

Figure 41 Two examples of elementary bifurcations occurring in a dissipative dynamical system. (a) Symmetric (pitchfork) bifurcation, in which a reference state x_0, loses its stability for $\lambda > 0$ and gives rise to two solution branches x_+, x_-, coalescing with x_0 at $\lambda = 0$. (b) Limit point bifurcation, whereby a stable branch x_+ and an unstable branch x_- collide and annihilate each other at $\mu = 0$.

solutions $x_\pm$ cannot be expanded in power series of the parameter. They therefore depend on λ in a *nonanalytic* fashion. This is the consequence, at the mathematical level, of the qualitative change of behavior introduced by the phenomenon of bifurcation.

Let us take an even simpler example. Consider the dissipative system

$$\frac{dx}{dt} = -x^2 + \mu \tag{3.12}$$

where the control parameter is now denoted by μ. The fixed points x_S are given by

$$x_\pm = \pm\sqrt{\mu} \tag{3.13}$$

and are plotted against μ in Figure 41(b). The time-dependent equation can again be solved exactly, and the solution shows that branch x_- is unstable, whereas branch x_+ is asymptotically stable. We notice that as μ decreases from positive values, the stable and unstable branches "collide" at $\mu = 0$ and subsequently are annihilated. For this reason we call $\mu = 0$ a *limit point*, or fold. Its presence signals the appearance once again of singularities, as illustrated by Eq. (3.13).

Let us now combine our two previous examples by considering the dissipative system

$$\frac{dx}{dt} = -x^3 + \lambda x + \mu \tag{3.14}$$

[We do not add the quadratic term of Eq. (3.12) since we can always eliminate it by a suitable change of variables.] We are led in this way to a problem involving two parameters, λ and μ. The fixed points are now given by the canonical form of the general cubic equation

$$-x_s^3 + \lambda x_s + \mu = 0$$

From elementary algebra we know that this equation can have up to three real solutions. Moreover, as the parameters vary the three solutions merge, and we are left with only one solution. We can determine curves in *parameter space* separating these two regimes, specifically,

$$-4\lambda^3 + 27\mu^2 = 0 \tag{3.15}$$

These curves are represented in Figure 42(a). The region of three real solutions ends at a point (the origin 0 in the figure), in which there is a singular dependence of λ on μ. This is known as *cusp singularity*.

Figures 42(b)(c) provide two different views of the dependence of the solutions on the parameters. In Figure 42(b) x_s is plotted against μ for fixed λ. The resulting S-shaped curve indicates the coexistence of multiple solutions for given parameter values. Moreover, two of these branches are simultaneously stable. The bistability region ends at the two limit points μ_1 and μ_2, in the vicinity of which we see the behavior shown in Figure 41(b). Under these conditions an increase of μ beyond μ_1 and up to μ_2, followed by a variation in the opposite direction, will lead to a *hysteresis cycle*, a phenomenon introduced in Section 1.4 in connection with the Belousov-Zhabotinski reaction.

In Figure 42(c) x_s is plotted against λ for fixed μ. We now obtain two disjoint curves: one, a, defined for all values of λ, and another, b, defined only for $\lambda \geqslant \bar{\lambda}$ and exhibiting a limit point singularity at $\bar{\lambda}$. For $\lambda < \bar{\lambda}$ only one stable solution is available, but for $\lambda > \bar{\lambda}$ we have bistability as before. Finally, Figure 42(d) combines the information in Figures 42(a)–(c) in a three-dimensional plot of the solution x_s versus the parameters λ and μ.

It is important to realize that for no nonvanishing $|\mu|$, however small, can the symmetric (pitchfork) bifurcation depicted in Figure 41(a) be observed. μ therefore acts like an *imperfection* that destroys bifurcation. On the other hand, the limit point, Figure 41(b) proves to be "robust," in the sense that it is recovered in both Figures 42(b) and (c). But if both λ and μ are varied simultaneously, there will always be a particular combination of values ($\mu = 0$, λ variable in our case) for which the bifurcation will be recovered, as the system will be able to traverse the cusp singularity [Figure 42(a)] in a symmetric fashion.

The above discussion illustrates a deep concept of great importance in the theory of dynamical systems, that of *structural stability*. It shows that certain phenomena, such as the pitchfork bifurcation above, occur only if the parameters present satisfy at least one equality. Inasmuch as in a physical system such a strict

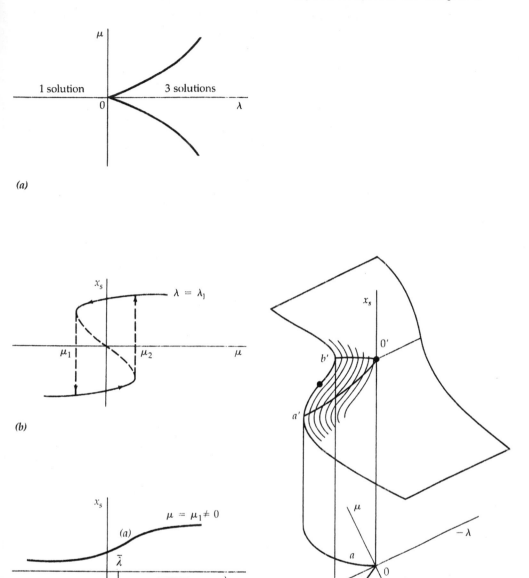

Figure 42 Effect of parameters in the bifurcation of steady state solutions of Eq. (3.14). (a) Curve of parameter space delimiting the region of existence of three real solutions. (b) Hysteretic behavior of the solution at fixed λ, as the parameter μ is varied. The limit point bifurcation remains robust. (c) Destruction of pitchfork bifurcation when the parameter μ, acting as an imperfection, is not identically zero. (d) Three-dimensional view combining the information of (a), (b), and (c).

requirement will be difficult to meet, we expect that these phenomena will disappear under slight changes of parameter values. We describe such phenomena as being *structurally unstable*. On the other hand there exist other phenomena, like the limit point bifurcation in Figure 41(b), that persist (even though they may be shifted) under changes of the control parameters affecting the structure of the evolution laws. These we call *structurally stable*. Note that according to this definition *all conservative systems are structurally unstable*, since the presence of small dissipative terms (like a slight friction in a pendulum) qualitatively alters the phase portrait by conferring the property of asymptotic stability to certain preferred solutions, the attractors.

From the point of view of the physical applications, what we have achieved here is to identify some minimal mathematical models reproducing one of the experimentally observed transition phenomena that accompany complex behavior—namely, the emergence of multiple simultaneously stable solutions describing the ability to switch and thus to perform regulatory tasks. In one-variable systems containing higher order nonlinearities more complex transition phenomena are possible. Catastrophe theory, a special branch of dynamical systems theory, aims at the classification of these phenomena and, in particular, at the determination of the relations among parameters that qualitatively delimit various types of behavior.

3.5 DISSIPATIVE SYSTEMS IN TWO-DIMENSIONAL PHASE SPACES: LIMIT CYCLES

The evolution of the one-variable systems considered in the preceding section took place in a one-dimensional phase space. In the terminology of Section 3.1, the only invariant sets that can exist in such a space are fixed points (steady states). The phase trajectories are simply straight half-lines converging to or diverging from the fixed points, depending on their stability properties, as illustrated in Figures 43(a) and (b) for the models of Eqs. (3.10) and (3.12), respectively.

In a two-dimensional phase space topological considerations clearly suggest a larger flexibility than is available in one-dimensional space, but our discussion in Section 3.1 above shows that very stringent conditions remain that limit the various possibilities. A two-variable system evolves in just such a limited two-dimensional phase space. We of course expect it to possess the invariant sets of one-variable systems, namely, fixed points. Moreover, inasmuch as conservative systems with one degree of freedom constitute a particular class of two-variable systems we expect to encounter, under certain conditions, the phase portraits of the integrable systems discussed in Section 3.3. In particular, fixed points could behave as saddles or centers, a continuum of closed trajectories surrounding the center could exist, and the separatrices of the saddle points could bend to form loops joining two unstable states, as in Figure 40. The question is whether, in

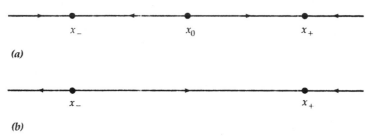

(a)

(b)

Figure 43 Phase space trajectories of a dissipative dynamical system involving one variable. (a) One unstable and two asymptotically stable solutions as in Figure 41(a). (b) One unstable and one asymptotically stable solution as in Figure 41(b). Initial conditions on the left of x_- lead to explosive behavior.

addition to the above, we can expect—as a consequence of the extra dimension of phase space—new possibilities specifically related to the dissipative character of the system.

Figure 44 describes a first possibility, related to the way the fixed point P is approached in time. In both cases depicted, P is asymptotically stable but the phase portraits in its vicinity are entirely different (this was clearly impossible in Figure 43). In Figure 44(a) the trajectories tend to P in a more or less radial way as time goes to infinity, corresponding to an exponential time decay of the perturbations around P. We call this configuration a (stable) *node*. In Figure 44(b) the trajectories spiral around P before finally converging to it. In the time dependence of the variables this will show up in the form of damped oscillations.

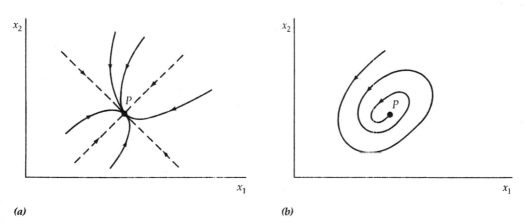

(a)   *(b)*

Figure 44 Possible phase space configurations around a fixed point P of a dissipative dynamical system involving two variables, in addition to the configurations (saddle and center) allowed for conservative systems. (a) Asymptotically stable node. Solutions evolve to P monotonically. (b) Asymptotically stable focus. Solutions evolve to P while performing damped oscillations. In both cases P is a zero dimensional attractor.

We call this configuration a (stable) *focus*. Note that these configurations but with all arrows reversed are also possible; in such a case we would be dealing with an *unstable* node or focus. Whatever their stability, nodes or foci can only arise in dissipative systems, since they clearly violate time reversal invariance.

The most interesting possibility is certainly that shown in Figure 45. In the domain of phase space considered, the system possesses a single fixed point, P, which for the parameter values chosen is unstable. Trajectories starting close to P cannot be attracted by it and tend to diverge. But in a physically reasonable situation the divergence cannot lead to infinity since that would imply that the system can absorb or yield an infinite amount of matter or energy. We therefore expect the trajectories to remain confined and still satisfy the topological requirements explained in Section 3.1 (no self-intersection, etc.). In a two-dimensional space the only remaining outcome is that the trajectories end up on a closed curve, denoted by C in Figure 45. We call such curves, first discovered by Poincaré, *limit cycles*. In the particular example chosen in Figure 45 the limit cycle is asymptotically stable. But once we realize the possibility of closed curves occurring at a finite distance from any other closed curve (contrary to conservative systems), other configurations can be envisaged. For instance, C can be orbitally unstable, and the repelled trajectories can tend to another limit cycle or to a stable fixed point. In any case, according to Section 3.1, once on a limit cycle the system follows the same succession of states forever; in other words, it shows periodic behavior in time. An asymptotically stable limit cycle thus constitutes a *periodic attractor*. By virtue of its asymptotic stability, it will be robust against perturbations and will thus be the natural architype for describing the reproducible rhythmic phenomena observed in nature, as surveyed in Chapter 1.

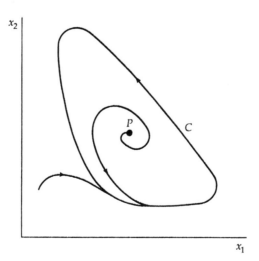

Figure 45 Periodic attractor: phase space trajectories evolve toward a limit cycle C, surrounding an unstable fixed point P.

Now let us study the mechanisms presiding over the formation of a limit cycle. To get a clear feeling of the minimal requirements and to bypass technicalities as much as possible, we first consider a model that has exact solutions. Imagine a conservative system of two variables (automatically integrable according to Section 3.3) such as the harmonic oscillator. In the action-angle variables the evolution is given by a set of equations similar to Eq. (3.8):

$$\frac{dr}{dt} = 0$$

$$\frac{d\varphi}{dt} = \omega$$

in which the action variable is now denoted by r for reasons that will become clear shortly. For a given value of r this system follows a closed trajectory like that in Figure 38, one that is reminiscent of a limit cycle but that differs in that it is part of a continuum of closed trajectories. We therefore modify the above system with a *dissipative perturbation* violating the time reversal symmetry. We choose the perturbation in such a way that a fixed point can also exist and the resulting limit cycle can be asymptotically stable, at least under certain conditions. Clearly it would not make much sense to apply a dissipative perturbation to the angle variable φ, which increases passively by 2π on each turn along the periodic trajectory. We therefore perturb the radial part by adding a nontrivial term on the righthand side of the equation. According to Section 3.4 [see Eq. (3.10)], a simple perturbation satisfying these requirements is a cubic. We thus write

$$\frac{dr}{dt} = \lambda r - r^3$$

$$\frac{d\varphi}{dt} = \omega$$

(3.16)

The first of these equations is identical to Eq. (3.10), with the additional requirement that the radius r be positive. The behavior of its solutions follows immediately from Figure 41(a) and is shown in Figure 46. For $\lambda < 0$ only the trivial solution $r = 0$ exists and is stable. In the phase space defined by the orthogonal coordinates

$$x_1 = r \cos \varphi, \qquad x_2 = r \sin \varphi$$

the trajectories will converge to the fixed point $x_1 = x_2 = 0$, which will behave as a focus [Fig. 46(b)] in view of the fact that the angle variable φ (phase) will vary according to

$$\varphi = \varphi_0 + \omega t$$

(3.17)

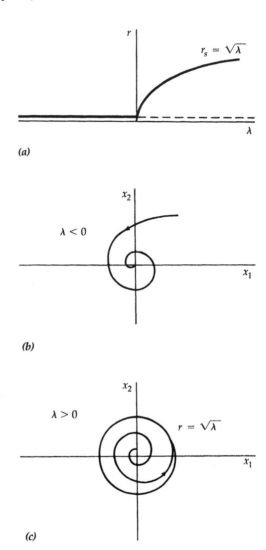

Figure 46 Hopf bifurcation illustrated on the simple model of Eqs. (3.16). (a) Radius of limit cycle as a function of the control parameter λ. (b), (c) Phase portraits for $\lambda < 0$ and $\lambda > 0$ respectively.

φ_0 being the initial phase. But when $\lambda > 0$ the first equation in (3.16) admits the nontrivial solution $r_s = \sqrt{\lambda}$, which is asymptotically stable, while the fixed point $r = 0$ is unstable. In the phase space this will be represented as in Figure 46(c): by trajectories spiraling out of the fixed point and converging to the closed curve $r_s = \sqrt{\lambda}$—the limit cycle. $\lambda = 0$ is therefore a point of bifurcation of a new type, known as a *Hopf bifurcation*. Contrary to the static bifurcations encountered in Section 3.4, which were all leading eventually to time-independent solutions, a

Hopf bifurcation is a dynamical phenomenon leading to time-periodic behavior. The amplitude of the solution grows smoothly from zero as the parameter λ moves away from its value at the bifurcation point, as seen clearly in Figure 46(a). Moreover, it depends on the distance from bifurcation in a nonanalytic fashion. The period of the oscillation is, however, generally finite at the bifurcation point, as long as ω is not identically zero in Eq. (3.16). In either case, the important point is that both period and amplitude are intrinsic to the system for they depend only on the values of λ and ω, which are built into the equations of evolution. This is strikingly different from conservative systems, in which the characteristics of the oscillations depend on the initial conditions. We thus have a simple mathematical model reproducing the essential features of a large class of rhythmic phenomena encountered in Chapter 1.

As long as r and φ in Eqs. (3.16) have the properties of a radial and angle variable, respectively, one can show that these equations do not suffer from the sensitivity to perturbations that characterized the solutions of Eq. (3.10). Thus, *a Hopf bifurcation is a structurally stable phenomenon.*

Two-variable dissipative systems can give rise to more intricate transition phenomena. However, most of the corresponding bifurcations are structurally unstable in the sense that they require that an equality between parameters be satisfied. In this respect it is worth noting that the theory of dynamical systems has achieved full classification of all structurally stable phase portraits that can occur in systems evolving in a two-dimensional phase space.

3.6 REDUCTION TO LOW-DIMENSIONAL SYSTEMS: ORDER PARAMETERS AND NORMAL FORMS

In the last two sections we saw that certain phenomena taking place in low-dimension phase spaces subsist when the structure of the equations of evolution is modified through a change of parameters. It is therefore expected that at least within the realm of one- and two-variable systems these phenomena are typical, in the sense that for large classes of such systems they should be observable in certain ranges of parameter values. Let us convince ourselves about this by considering a few simple models that are not abstract mathematical examples but instead derive from the laws of physics and chemistry.

Pitchfork and limit point bifurcations

These phenomena are nicely illustrated by models from chemical kinetics. As discussed in Sections 1.4 and 2.4, a chemical reaction usually involves collisions between molecules, and to a good approximation the frequency of collision is proportional to the product of the concentrations of the species considered. This generates polynomial nonlinearities, which were at the basis of the results derived in Section 3.4.

A useful visualization of these possibilities is provided by the one-variable autocatalytic model introduced in Eq. (2.15). As shown in Eq. (2.17), the fixed points are the solution of a cubic equation. At equilibrium this equation admits only one solution, but far from equilibrium the alternative change of the signs of the various terms suggests up to three real solutions. Now, from algebra, any such cubic expression can be transformed to a canonical form equivalent to Eq. (3.14) by a simple linear change of variables. It follows that the model, Eq. (2.15), which is compatible with the requirements imposed by thermodynamics and chemical kinetics, gives rise to all the static bifurcation phenomena discussed in Section 3.4.

Hopf bifurcation and limit cycles

Here too chemical kinetics provides nice illustrative examples. One of the most extensively studied is the trimolecular model or *brusselator*. It involves two coupled intermediates and one autocatalytic step, and for simplicity it is usually considered in the limit of irreversible reactions:

$$A \rightarrow X$$
$$B + X \rightarrow Y + D$$
$$2X + Y \rightarrow 3X$$
$$X \rightarrow F$$

(3.18)

It is also assumed that the concentrations of A and B are controlled from outside. In the limit of a spatially uniform (well-stirred) system the rate equations become (in suitably rescaled variables)

$$\frac{dX}{dt} = A - BX + X^2Y - X$$

(3.19)

$$\frac{dY}{dt} = BX - X^2Y$$

By equating the time derivatives to zero, we find that the system possesses a unique fixed point, $X_s = A$ and $Y_s = B/A$. Moreover, by standard analytical methods we can show that this point is a stable focus, as shown in Figure 44(b), if B is slightly less than $A^2 + 1$, and an unstable focus if B is slightly larger than $A^2 + 1$. In this latter range a limit cycle solution, depicted in Figure 47, exists. Its characteristics are exactly those described for the abstract mathematical model used in Section 3.5. In other words, the critical value

$$B_c = A^2 + 1$$

(3.20)

is a Hopf bifurcation point. In fact, the analogy between Eq. (3.19) and the abstract mathematical model goes much further, but we postpone this discussion for a moment.

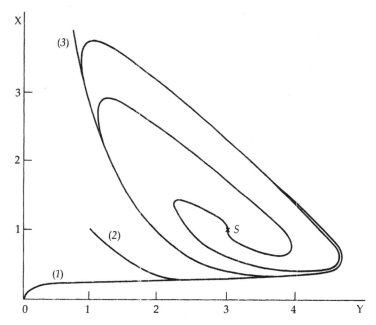

Figure 47 Numerically computed limit cycle for the brusselator model, Eq. (3.19). Parameter values: $A = 1$, $B = 3$.

Electrical circuits comprise another class of systems of two variables giving rise to limit cycles. Here the nonlinearities arise from the presence of *active elements* (electronic tubes, transistors, and so on) that induce nonlinear current-voltage characteristics. A classic example is the inductive coupling of a triode and a resonant circuit, first studied extensively by the Dutch scientist Balthazar van der Pol. The application of the well-known laws of conservation of charge for circuits leads to a closed equation for the voltage x across the inductance of the resonant circuit, which in dimensionless variables reads:

$$\frac{d^2x}{dt^2} - \varepsilon(1 - x^2)\frac{dx}{dt} + x = 0 \tag{3.21}$$

where the parameter ε is related to the properties of both the triode and the resonant circuit. The analysis of this equation in the classic work of van der Pol showed the existence of a sustained oscillation of the limit cycle type. In electronics the emergence of such oscillations is a well-known, if sometimes undesirable, phenomenon referred to as self-excitation.

But the impact of the bifurcations described in Sections 3.4 and 3.5 goes far beyond their applicability to one- or two-variable systems. In most physical problems we are faced at the outset with a host of variables, for example, about 10 in the Belousov-Zhabotinski reaction. Consequently, if these bifurcations were limited to the rather exceptional systems discussed above they would constitute interesting curiosities rather than powerful tools of analysis of natural systems.

Fortunately this is not so. There exist two quite general (but not completely unrelated) mechanisms by which the formulation of pitchfork, limit point, and Hopf bifurcations outlined previously can be applied to extremely wide classes of systems independently of the numbers of variables and the degree of nonlinearities involved.

Let us begin with the conceptually simpler mechanism, which should immediately evoke familiar concepts to anyone with training in chemistry. In many cases the dynamics of a physico-chemical system gives rise to a variety of widely separated time scales due to order of magnitude differences in the values of parameters and/or the state variables. For instance, in chemistry a catalytic reaction under laboratory conditions usually involves catalyst concentrations that are much less than the initial or final product concentrations, and rates that are much higher than the intrinsic rates in the absence of catalyst. As a result, some intermediate steps involving catalytic complexes proceed very quickly. In combustion the activation energies of some of the exothermic reactions are very high, so these reactions proceed, at least in the early stages, much more slowly than do energy and momentum transport. Similar examples can be found in biology, optics, and other fields.

Intuitively, we expect that during a time interval in which the "slow variables" hardly change, the "fast variables" will evolve a long way until they reach the level expected from the steady state solution of their evolution equations. But there will be one major difference: The values of the slow variables that the fast ones will "see" at this stage will not be their final steady state levels, but rather the instantaneous values predicted by the slow variable equations. Denoting the fast and slow variables by Y and X, respectively, we have in this quasi steady state a relation of the form

$$Y = h(X) \tag{3.22a}$$

playing essentially the role of an *equation of state*. Substituting this expression in the equations for the slow variables, in which both X and Y intervene, we obtain

$$\frac{dX}{dt} = F(X, Y) = F[X, h(X)]$$
$$= f(X) \tag{3.22b}$$

which constitutes a closed set of equations for X. We have therefore succeeded in considerably reducing the number of variables involved by eliminating the fast ones. This is analogous to the ideas behind the rate-limiting step and the quasi-steady-state approximation familiar in chemistry and, most particularly, in enzyme kinetics. If the number of the remaining slow variables is one or two, which is quite frequently the case, we can apply the results of Sections 3.4 and 3.5 to systems that are much more general than the abstract mathematical models we considered there. An explicit example will be treated in Section 6.3.

In our discussion we have appealed repeatedly to intuition. But thanks to a basic theorem demonstrated by the Soviet mathematician Andrei Tikhonov, in which the detailed conditions that the rate functions and the parameters must satisfy are spelled out explicitly, we can express all our major conclusions in rigorous terms.

Let us now consider the second, much subtler, mechanism allowing the reduction of an initial multivariable system to few variables. Suppose that we have a set of evolution equations of the form of Eq. (3.1), in which the number n of variables involved can be very high. By a standard method known as *linear stability analysis*, we can determine the parameter values λ for which a certain reference state $\{X_{js}\}$ switches from asymptotic stability to instability. Readers interested in technical details will find a compact presentation of this method in Appendix 1, along with illustrations on simple models; here we outline the main ideas.

As discussed in Section 2.6, stability is essentially determined by the response of the system to perturbations or fluctuations acting on a reference state. It is therefore natural to cast the dynamical laws, Eq. (3.1), in a form in which the perturbations appear explicitly. Setting

$$X_i(t) = X_{is} + x_i(t) \tag{3.23}$$

substituting into Eq. (3.1), and taking into account that X_{is} is also a solution of these equations, we arrive at

$$\frac{dx_i}{dt} = F_i(\{X_{is} + x_i\}, \lambda) - F_i(\{X_{is}\}, \lambda)$$

These equations are homogeneous in the sense that the righthand side vanishes if all $x_i = 0$. To get a more transparent form of this homogeneous system, we expand $F_i(\{X_{is} + x_i\}, \lambda)$ around $\{X_{is}\}$ and write out explicitly the part of the result that is linear in $\{x_j\}$, plus a nonlinear correction whose structure need not be specified at this stage:

$$\frac{dx_i}{dt} = \sum_j L_{ij}(\lambda)x_j + h_i(\{x_j\}, \lambda) \qquad i = 1, \ldots, n \tag{3.24}$$

L_{ij} are the coefficients of the linear part and h_i the nonlinear contributions. The set of L_{ij} defines an *operator* ($n \times n$ matrix in our case), depending on the reference state X_s and on the parameters λ.

Now a basic result of the theory, discussed in more detail in Appendix 1, establishes that the properties of asymptotic stability or instability of the reference state $X = X_s$ (or $x = 0$) of system (3.24) are identical to those obtained from its linearized version:

$$\frac{dx_i}{dt} = \sum_j L_{ij}(\lambda)x_j \qquad i = 1, \ldots, n \qquad (3.25)$$

Stability reduces in this way to a linear problem that is soluble by methods of elementary calculus. It is only in the borderline case in which $X = X_s$ (or $x = 0$) is Lyapounov stable but not asymptotically stable that the linearization might be inadmissible.

Figure 48 summarizes the typical outcome of a stability analysis carried out according to this procedure. What is achieved is the computation of the rate of growth γ of the perturbations as a function of one (or several) control parameter(s). If $\gamma < 0$ (as happens in Figure 48, branch 1 when $\lambda < \lambda_c$) the reference state is asymptotically stable; if $\gamma > 0$ ($\lambda > \lambda_c$ for branch 1) it is unstable. At $\lambda = \lambda_c$ there is a state of *marginal stability*, the frontier between asymptotic stability and instability.

In general a multivariable system gives rise to a whole spectrum of γ, just as a crystal has a multitude of vibration modes. We will therefore have several γ versus λ curves in Figure 48. Suppose first that of all these curves only one,

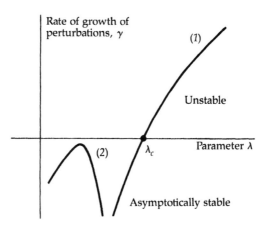

Figure 48 Rate of growth, γ, of perturbations as a function of the control parameter λ deduced from linear stability analysis [cf. Eq. (3.25)]. Curve (1) The reference state is asymptotically stable for $\lambda < \lambda_c$ and becomes unstable for $\lambda > \lambda_c$, where λ_c is the critical value of marginal stability. Curve (2) The reference state remains asymptotically stable for all values of λ.

curve *1* in the figure, crosses the λ axis, while all others are below the axis. Under well-defined mild conditions, discussed further in Appendix 2, we can then show that at $\lambda = \lambda_c$ *a bifurcation of new branches of solutions takes place*. Two cases can be distinguished:

1. If at $\lambda = \lambda_c$ the perturbations are nonoscillatory [as happens, for instance, in Figure 44(a)] the bifurcating branches will correspond to steady-state solutions
2. If at $\lambda = \lambda_c$ the perturbations are oscillatory (as in Figure 45), the bifurcating branches will correspond to time-periodic solutions in the form of limit cycles

In either case, for the description of the bifurcating branches as well as for transient behavior, a suitable set of quantities can be defined which obey a closed set of equations if the parameters lie close to their critical values λ_c. In case 1 there is only one such quantity, and it measures the amplitude of the bifurcating branches. In case 2 there are two such quantities characterizing the amplitude and the phase of the oscillation. Therefore the original dynamics is effectively decoupled into a single equation or a pair of equations giving information on bifurcation, and $n - 1$ or $n - 2$ equations that are essentially irrelevant as far as bifurcation is concerned. We call the quantities that satisfy the bifurcation equations *order parameters*.

A most surprising point is that the bifurcation equations turn out to have exactly the same form as the abstract mathematical models of sections 3.4–3.5 that gave rise to pitchfork and Hopf bifurcation. In other words, any dynamics satisfying the conditions of either case 1 or case 2 above can be cast in a universal, *normal form* close to the bifurcation point. This powerful result was in fact known to Poincaré, but it was the American mathematician George Birkhoff who developed it for the study of bifurcations in conservative systems. For dissipative systems the idea of reduction to a few order parameters is frequently associated with the name of the Soviet physicist Lev Landau, in connection with his theory of phase transitions. In the mathematical literature it is frequently referred to as the Lyapounov-Schmidt procedure, and more recently as the *center manifold theory*. A compact derivation of the bifurcation equations is given in Appendix 2.

More intricate situations can also be envisaged in which several branches cross the λ axis in Figure 48. This leads to interaction between bifurcating solutions that generates *secondary, tertiary,* or even higher order bifurcation phenomena. The above approach to reduction will still apply, in the sense that the part of the dynamics that gives information on the bifurcating branches takes place in a phase space of reduced dimensionality. However, the explicit construction of the normal form becomes much more involved and its universality can no longer be guaranteed.

In summary, we can assert that the static and dynamic bifurcations surveyed in Sections 3.4 and 3.5 are authentic *archetypes of complexity*, in the sense that they are present in large classes of systems under quite different conditions. In the next section we will see that despite their generality, these bifurcations constitute only a part of the amazing repertoire of behaviors of dynamical systems.

3.7 PHASE SPACE REVISITED: TOPOLOGICAL MANIFOLDS AND FRACTALS

We have seen that in all systems where dynamics is reducible to a one- or two dimensional phase space, the behavior is severely limited by the topological constraints imposed on any low-dimensional motion. In particular, the only possible attractors turn out to be fixed points and limit cycles.

We will now be interested in the qualitative properties of dynamical systems evolving in phase spaces of dimension greater than two. At the end of Section 3.1 we stressed that the freedom afforded by additional dimensions was likely to be at the origin of completely new behaviors. Independently of these new aspects, we expect to recover all characteristics of one- and two-dimensional systems, and in particular the fixed point and limit cycle attractors. Naturally, the topological structure of the trajectories around the attractors will be more involved, but over a sufficiently long time the higher dimensional systems will tend to the same behavior as one- and two-dimensional systems. Let us therefore leave this problem aside and turn our attention to the genuinely new possibilities.

In order to get as clear a picture as possible, we will discuss systems evolving in three-dimensional phase space. In our exploratory analysis we want to benefit from the experience we have already acquired in the handling of two-dimensional problems—let us try to figure out a good strategy. One possibility is suggested by the story of the shadows on the cave walls in Plato's *Republic*: Look at the "shadow" of the full object, and try to reconstitute it from that two-dimensional picture. Plato admits that this is a very difficult enterprise, and asserts that only philosophy can bring the human mind close to this achievement. In more modern terms, we know that no one particular projection is in one-to-one correspondence with the object that generates it. Several projections on differently oriented surfaces are necessary in order to fully describe the object. But this automatically brings us to the original multidimensional space, which is precisely what we wanted to avoid.

Fortunately, a second possibility suggested by another great genius, Henri Poincaré, is open to us. We consider a plane that cuts the phase trajectories transversely, as in Figure 49, and study the succession of points P_n at which the trajectories intersect the plane. Depending on the object that will attract the P_n's for long times, we will be able to infer the attractor embedded in the three-dimensional phase space, since in actual fact we will dispose of a section of this attractor.

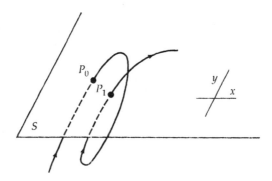

Figure 49 The Poincaré surface of section. The intersections of the trajectories of a dynamical system evolving continuously in time with a surface S transverse to the trajectory are studied. The discrete time dynamics describing the position of the successive intersection points $P_0, P_1, \ldots$ on S gives extensive information on the underlying continuous flow.

S is referred to as the Poincaré surface of section, and the dynamics of the P_n's as the *Poincaré map*. Note that this dynamics is in fact a *recurrence* in which the time intervenes in a discrete fashion, since the time intervals between successive intersections will be finite. By introducing an appropriate coordinate system in the plane S, we can write this recurrence in the form

$$x_{n+1} = f(x_n, y_n)$$
$$y_{n+1} = g(x_n, y_n)$$
(3.26)

Now that the dynamics has been reduced to two-dimensions, we can apply our previous results and make plausible conjectures. Let us attempt a preliminary classification of the various possibilities.

Periodic attractors: cycles of order k

Figure 50(a) suggests the most obvious possibility: The intersection points $P_0, P_1, \ldots$ converge to a fixed point P. This obviously means that the attractor, if bounded, is a limit cycle, C_1. In the recurrence relation, Eq. (3.26), the existence of P will show up as the possibility of satisfying simultaneously the equations $x_{n+1} = x_n, y_{n+1} = x_n$ or equivalently $x = f(x, y), y = g(x, y)$.

Other possibilities are allowed. Instead of the cycle of order one just defined, we can require that $x_{n+1}, x_{n+2}, \ldots$ remain different from x_n (similarly for y) until an iteration k is reached for which $x_{n+k} = x_n, y_{n+k} = y_n$. We call this a cycle of order k. Figure 50(b) describes how a cycle of order two should look. In the Poincaré surface the attractor is a set of two points, P and P', visited successively

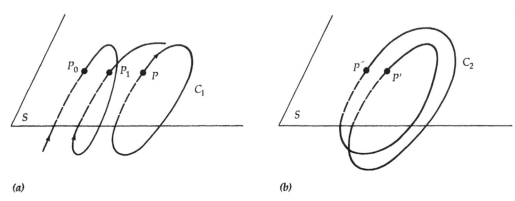

(a)   (b)

Figure 50 Convergence of the trajectories of a continuous-time dynamical system toward a limit cycle C_1 and a limit cycle C_2 whose period is twice the period of C_1. In the Poincaré surface of section this is manifested by the convergence of the sequence $P_0, P_1, \ldots$ respectively to a stable fixed point P and to a pair of points P, P'.

by the system, whereas the corresponding object in the three-dimensional space is a curve C_2 that gets twisted before closing to itself. This would be impossible in a two-dimensional phase space as it would imply self-intersection of the trajectories. Clearly this is the natural model for describing periodic phenomena whose period changes abruptly as one of the parameters is varied. Such transitions have been abundantly observed experimentally in fluid mechanics, chemistry, and optics and are discussed further in Section 3.10.

Note that, as in Section 3.5, the fixed points of the Poincaré map can also be classified as saddles, nodes, and so on.

Quasi-periodic attractors: invariant tori

Another topologically permissible possibility is suggested by Figure 51. The intersection points $P_0, P_1, \ldots$ converge to a closed curve C in the Poincaré surface S. The corresponding attracting object in the three-dimensional space is therefore a two-dimensional surface, and if we require on physical grounds that the trajectories do not go to infinity, the surface will have to be bounded. In addition, its structure must allow the trajectories to evolve without intersection as required by the uniqueness theorem described in Section 3.1. This rules out the sphere and leaves us with the *torus*, encountered in Section 3.3. As explained there, the motion on a toroidal surface is generally quasi-periodic, with two independent incommensurate frequencies. The new element here is that the torus is attracting, contrary to the case of conservative systems discussed in Section 3.3 in which there was a continuum of tori surrounding a periodic orbit.

In other words, an attracting invariant torus is the natural model of reproducible quasi-periodic phenomena observed in nature.

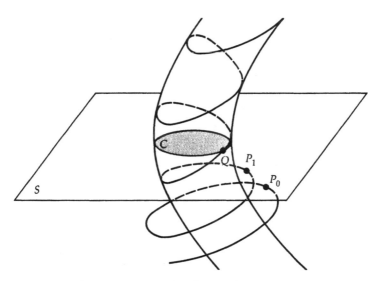

Figure 51 Quasi-periodic attractor in the form of a two-dimensional toroidal surface T^2, embedded in a three-dimensional phase space. During the transient stage of evolution the trajectory intersects the Poincaré surface of section S on points $P_0, P_1, \ldots$ before converging to the surface of the torus. Subsequently it keeps winding indefinitely on this surface. At each passage through S the trajectory gives rise to an intersection point such as Q_1. The set of all traces of successive passages constitutes the curve C, which is the attractor on the Poincaré surface and to which the sequence of transient points $P_0, P_1, \ldots$ converges.

Nonperiodic attractors: fractals

Could attractors other than fixed points, closed curves, or bounded two-dimensional surfaces exist in a three-dimensional phase space? We first note that an attractor must necessarily have a lower dimensionality than the phase space itself, since it represents a set of states that remains invariant under the dynamics. The question therefore amounts to asking whether there exist objects whose dimensionality is between that of a point and a line, of a line and a surface, or even of a surface and a volume. Such objects, if they exist, will not be points, nor curves, nor surfaces, nor—phrased more generally—topological manifolds. The French mathematician Benoit Mandelbrot coined the name *fractals* for them.

The theory of numbers, which is generally regarded as one of the deepest areas of mathematics, allows us to foresee certain possibilities, since the question raised above can be reduced to the following: Do sets exist that are intermediate between a denumerable set such as the natural numbers, and a continuum such as the points on a line? The answer to this question is yes, and it is intimately related with the notion of infinity. It was provided by the German mathematician Georg Cantor. We cannot go into the subtleties of the mathematical reasoning here, so we will illustrate the result with a particular example known as the *Cantor set*, shown in Figure 52.

Consider the closed segment [0, 1]. We divide it in three equal parts and delete the middle one without its boundaries. We divide each of the remaining segments in three equal parts and again delete the open middle one. And so forth. By construction, the set of points obtained in the limit is an infinite, nondenumerable set of nonconnected points. It has no intrinsic length (it is of *measure zero* in more technical terms), since the length of its complement is equal to one: $\frac{1}{3} + \frac{2}{9} + \frac{4}{27} + \cdots = 1$. Surprisingly, however, its dimension is nonzero! Let us see how this can happen.

How do we express in abstract terms that the piece of paper on which we write is two-dimensional, or that the space in which we move is three-dimensional? Call D whatever the set of interest might be, and try to cover it by segments, or squares, or more generally, hypercurbes of side ε. Observe then that the dimension d of all well-known "topological" manifolds derives from the following relation:

$$d = \lim_{\varepsilon \to 0} \frac{\ln N_\varepsilon}{\ln 1/\varepsilon} \qquad (3.27)$$

The concept is illustrated in Figure 53. We choose cells of characteristic size ε; count the minimum number of cells N necessary to cover D; divide the logarithm of N by the logarithm of $1/\varepsilon$; take the limit of very small ε; and we obtain d, the dimensionality of D.

Let us apply this for the plane, specifically a square of side 1. We need $(1/\varepsilon)^2$ squares of side ε to cover this space, hence $d = \ln (1/\varepsilon)^2/\ln (1/\varepsilon) = 2 \ln(1/\varepsilon)/\ln (1/\varepsilon)$; i.e., $d = 2$ as expected. But what do we obtain if we extend this concept to the Cantor set?

In the first partition of the interval, two segments of length 1/3 suffice to describe all the subsets present. In the second partition, however, we need four segments of length 1/9, and more generally, in the nth partition we need 2^n segments of length $1/3^n$. Thus $\ln N_\varepsilon/\ln (1/\varepsilon) = \ln 2^n/\ln 3^n = (n \ln 2)/(n \ln 3)$, or

$$d_{\text{Cantor}} = \frac{\ln 2}{\ln 3} \simeq 0.63 \qquad (3.28)$$

$$0 \qquad \frac{1}{9} \qquad \frac{2}{9} \qquad \frac{1}{3} \qquad \frac{2}{3} \qquad \frac{7}{9} \qquad \frac{8}{9} \qquad 1$$

Figure 52 Construction of the Cantor set. Shaded sections are the center portions eliminated in successive partitions of the line segments into thirds.

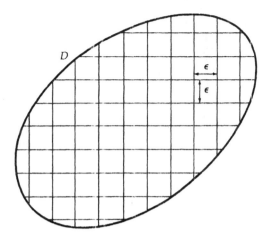

Figure 53 Illustration of the concept of dimensionality, d, of a set D. The minimum number of hypercubes of size ε necessary to cover the set completely is counted for decreasing values of the size ε, and the limit of Eq. (3.27) is evaluated.

The Cantor set is therefore in this respect intermediate between a point ($d = 0$) and a line ($d = 1$): it is a *fractal object*.

Having established the existence of such sets we can now imagine attractors in a three-dimensional phase space that are manifested in the Poincaré surface of section by a nondenumerable infinity of highly folded segments, visited successively by the phase space trajectory in a nonreproducible fashion as time follows its course. The corresponding object in three-dimensional space consists of an infinity of sheets such that a section taken transverse to the sheets is a Cantor set. It is natural to expect that such objects, frequently referred to as *strange attractors*, will constitute the models describing the onset of chaotic, turbulentlike behavior.

We have centered our discussion on attractors, which can exist only in dissipative systems, but all the arguments evoked can be used to describe the invariant sets of a conservative system as well. Our task in the next three sections will be to convince ourselves that a number of possibilities which at this stage are mere speculations motivated by topological considerations, actually correspond to physical reality.

3.8 NONINTEGRABLE CONSERVATIVE SYSTEMS: THE NEW MECHANICS

The integrable conservative systems considered in Section 3.3 constitute a rather exceptional class. Let us think for a moment about the movement of planets around the sun. The mass of the planets is about a thousand times less than the mass of the sun, so at first approximation we can neglect their interaction and take into account only their attraction by the sun. In this way we obtain the *two-body problem*, which provides a classic example of an integrable (and exactly soluble) system giving rise to a periodic motion for each individual planet, and a quasi-periodic motion for the entire planetary system.

The interaction between planets, small as it may be, is nevertheless always present and will tend to perturb the keplerian trajectories. The question is whether this perturbation induces only slight quantitative adjustments or on the contrary, modifies the qualitative features of the motion. This is the basic problem facing celestial mechanics; very similar questions can be raised for a variety of other mechanical systems.

Clearly, as long as one is interested in time scales of the order of the inverse of the strength of the perturbation (these would be in the 1000-year range for our planetary system), the effects induced by the latter can be handled straightforwardly. It suffices to look for approximate solutions of the full equations by adding a small correction, proportional to the strength of the perturbation, to the zero-order term representing the integrable system of reference. This can be done by standard methods. But if we look for solutions valid for very long times and, in particular, if we want to understand the qualitative behavior of the exact solutions, we are immediately faced with a number of formidable difficulties. In our example of the planetary system these would arise in a time scale in the range of 1 billion years, and the question at issue could be whether the planets will escape from the system, crash on the sun, or hit each other. Less dramatic but technically similar questions arise in regard to the motion of fast charged particles in intense magnetic fields like those prevailing in the storage rings of giant particle accelerators, or in regard to the orbital stability of a space craft.

The principal manifestation of such difficulties is the presence of *divergencies*. When we use traditional perturbation methods to evaluate the correction to the motion of the integrable system, we find that it becomes very large or even tends to infinity under certain conditions. The most ubiquitous such condition is the occurrence of *resonance*, that is, the case where the frequencies of the integrable reference system are commensurate in the sense of $k_1\omega_1 + k_2\omega_2 + \cdots = 0$ (see Section 3.3). Such catastrophes signal the breakdown of the method used and suggest strongly that there should be a qualitative change in the behavior of the system under these conditions.

Let us try to understand, on intuitive grounds, the two aspects of the difficulty: First, why are resonances abundant rather than exceptional; and second, why are they "dangerous"?

The theory of numbers tells us that the irrationals constitute the vast majority of real numbers since, for one thing, they constitute a nondenumerable set (contrary to the rationals) and they carry the whole "length" or "measure" of any segment on the real line. Thus [see Eqs. (3.8)] in most of the points of the phase space the action variables $\{I_j\}$ will be such that the frequencies $\{\omega_j\}$ realized by the system will be irrationally related and hence incommensurate. The difficulty of resonance will simply not arise, and as a rule the motion will take place on *the nonresonant tori* [Figure 39(b)] generated by such sets of irrationally related frequencies. But things are not so simple. Even if they are "exceptional" in some well-defined sense, rational numbers remain sufficiently abundant (in mathematical terminology we speak of rationals being *dense* in the set of real numbers), since in any prescribed vicinity of any real number there always exist rational numbers. Thus, any *finite volume* of phase space, however small, will contain an infinite number or representative points in which the resonance condition is satisfied. In particular, resonances are expected to play an important role in the evolution of the underlying Gibbs ensemble (Section 3.2) and through it, in a statistical description of the system.

Let us now see why resonances are likely to be "dangerous." Consider again the resonance condition in connection with the second of Eqs. (3.8). What is really implied is that there exists a particular combination of angle variables, $\varphi = k_1 \varphi_1 + \cdots + k_n \varphi_n$, that remains invariant during the motion and might as well be eliminated from the description. In other words, our initial choice of n independent angles was misleading since only $n - 1$ of them are really independent. In more technical terms, our n-dimensional torus shrinks into an $(n - 1)$-dimensional one. We call such lower dimension invariant surfaces *resonant tori*. For two degrees of freedom this would mean that the invariant surface is covered by a family of closed (one-dimensional) orbits describing a periodic instead of a quasi-periodic motion. A single one of these orbits will generate on a Poincaré surface of section S—for example, curve C in Figure 54—a number of fixed points (P_1, P_2 in the

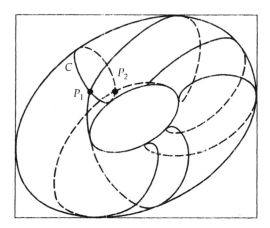

Figure 54 Fixed points P_1, P_2 generated on the intersection C between the Poincaré surface of section and a resonant torus as a particular periodic trajectory winds on this torus. Any other member of the continuum of periodic trajectories will generate an equal number of fixed points on the Poincaré surface: the whole curve C will thus be made up of fixed points. This pathological situation is expected to be disrupted by the action of a typical (nonintegrable) perturbation.

figure) equal to the number of turns the orbit makes in the axial direction (transverse to the surface of section) before closing.

Any other member of the family will of course generate another set of an equal number of fixed points lying on S. Thus, the whole curve C will be made up of fixed points. It is difficult to avoid the feeling that this should be an exceptional situation comparable, say, to having three randomly chosen points lie on a straight line. It is therefore expected that the most typical (nonintegrable) perturbations to which the integrable system may be subjected will disrupt this configuration of trajectories in phase space. This will be manifested through the fact that the correction to the unperturbed motion will be as important as the motion itself.

The search for a solution suitable for handling the difficulties raised by nonintegrable systems has led to a major scientific revolution, as a result of which a new mechanics has emerged since the 1950s. Many of the ingredients of this development were known to Poincaré and Birkhoff, but the Soviet mathematician Andrei Kolmogorov is to be regarded as the principal instigator of this breakthrough. Important contributions have subsequently been made by Vladimir Arnold and Jürgen Moser among others. Let us briefly review the highlights.

Perturbation of quasi-periodic motions

The arguments used earlier to suggest that resonant tori should be sensitive to small nonintegrable perturbations prompt us to expect that in the nonresonant case the motion should remain robust. In other words, typically, quasi-periodic behavior should not be altered greatly by a slight perturbation. This important result has been proven rigorously by Kolmogorov, who in addition was able to characterize in a sharp way those quasi-periodic motions that persist despite perturbation. Specifically, it appears that for the quasi-periodic motion to persist the frequencies must be "sufficiently" irrationally related in the following sense (illustrated here for two degrees of freedom): the ratio ω_1/ω_2 must be sufficiently different from a rational value,

$$\left| \frac{\omega_1}{\omega_2} - \frac{r}{s} \right| > \frac{c}{s^\nu} \qquad \text{for all integers } r \text{ and } s \qquad (3.29)$$

where c is a small number vanishing in the absence of the perturbation, and ν is sufficiently large (say $\nu = 2.5$). It can be shown that the quasi-periodic motions allowed by Eq. (3.29) constitute the majority of the motions realized in phase space.

In short, we expect that in vast regions of phase space a nonintegrable system close to an integrable system should show a behavior similar to Figure 39, namely, stable motions surrounding periodic orbits.

Perturbation of periodic motions

It is convenient to approach this problem in the language of the Poincaré surface of section S (Figure 54) and to inquire first about the fate of the continuum of fixed points on S under the effect of the perturbation. Let s be the winding number of the orbits. It can then be shown (see Appendix 3 for a succinct explanation) that the perturbation gives rise to $2ks$ fixed points (k is a certain integer), of which ks are elliptic and ks are hyperbolic (saddles). The elliptic fixed points correspond to stable periodic orbits, and each is surrounded by (non-resonant) invariant tori showing up in the Poincaré map as closed curves. The corresponding part of phase space is frequently referred to as an "island"; region a in Figure 55 is just such an island.

The situation is completely different for the hyperbolic fixed points. Because of the instability of the corresponding periodic orbits, an extremely intricate structure develops in their neighborhood, a structure that is essentially due to the peculiar configuration assumed by the separatrices of the orbits, as in region b of Figure 55. Specifically, in the Poincaré map these separatrices bend and intersect in an infinity of points known as homoclinic points (see Appendix 3).

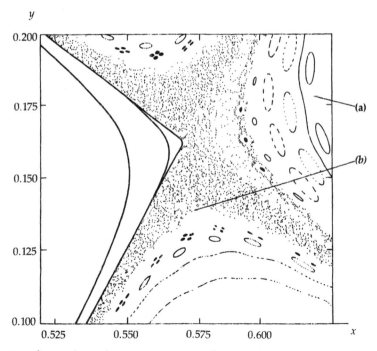

Figure 55 Regular and stochastic trajectories for a nonintegrable hamiltonian system visualized on a Poincaré surface. Region (a), stable motion near an elliptic fixed point. Region (b), unstable motion near a hyperbolic fixed point. [From M. Hénon, *Q. Appl. Math.*, **27**, 291 (1969)]

In these regions there appears a regime of highly *irregular motion*: the trajectories are very sensitive to slight changes in the initial conditions and wander in an erratic manner over large regions of phase space. In the Poincaré map this shows up as a cloud of points that seem to be randomly disposed, even though the trajectories are still absolutely deterministic. This chaotic behavior suggests that there may be a deep relation between nonintegrable conservative systems and the stochastic behavior we usually associate with irreversibility and dissipative systems. This most important point will be taken up in detail in Chapter 5.

Note that the overall structure is repeated on all scales about all elliptic fixed points that are generated on the Poincaré surface from the various resonant orbits. This remarkable self-similarity shows the enormous complexity attained in quite ordinary-looking problems, such as that of two weakly interacting degrees of freedom. A great deal of numerical work aiming to clarify this behavior has been carried out on simple models. For small values of the perturbation the calculations corroborate the general picture given in Figure 55. But as the energy of the system increases we observe that the islands tend to retreat, and the chaotic trajectories progressively invade ever larger portions of phase space. It is difficult to draw qualitative conclusions from such observations, since this regime is hardly accessible to analytical calculations or even to general topological considerations. Still, it is possible to introduce a measure characterizing the instability of the motion and the sensitivity to the initial conditions. This is provided by the *Lyapounov exponents*, which describe the mean rate of exponential divergence of two initially close trajectories. Lyapounov exponents constitute the generalization of the concepts of linear stability theory, introduced in Section 3.6, to chaotic systems. For more details refer to Appendix 1.

The complicated structure of regular and irregular trajectories discussed above is not an exclusive feature of nonintegrable hamiltonian systems. It is the typical behavior for other wide classes of conservative systems, among which the so-called Kolmogorov systems (K-systems) play an important role. We defer our discussion of K-systems to Chapter 5, as they will constitute our principal tool for illustrating the passage from conservative to dissipative dynamics. A very instructive approach to the study of unstable dynamical systems is through discrete time mappings, because of their intimate relation via the Poincaré surface of section [see Eq. (3.26)] to continuous-time systems. As a matter of fact, aside from some rare exceptions, discrete mappings are the only dynamical systems known on which we can establish explicitly and rigorously the emergence of irregular motions displaying chaotic behavior. A famous example invented by the American mathematician Stephen Smale is the *horseshoe map*, which can be formulated either as an abstract conservative or as a dissipative dynamical system. Although no known hamiltonian system can be reduced explicitly to this map, we describe it below and subsequently will use it as an architype of this new behavior, in the same sense that we have used cycles and tori as archetypes of simpler kinds of behavior.

3.9 A MODEL OF UNSTABLE MOTION: THE HORSESHOE

We consider a map, T, of the plane into itself which acts on a square R in the manner shown in Figure 56. The square is contracted in the vertical direction and expanded horizontally to form a "bar" that is then bent into a horseshoe shape, such that the arc portion and the ends of the horseshoe project outside the area of the original square. In this way R is not mapped onto itself, only the bands I and J of the straight legs of the horseshoe are contained in R. Their "pre-

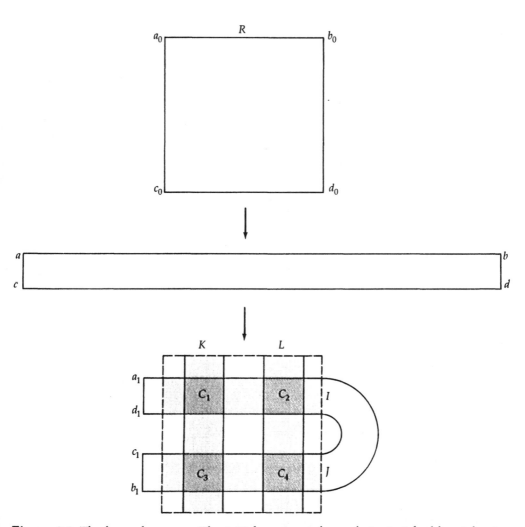

Figure 56 The horseshoe map. The initial square $a_0\ b_0\ c_0\ d_0$ is stretched by at least a factor of two in the horizontal direction and compressed by a factor larger than two in the vertical direction and subsequently bent to form a horseshoe. Squares C_1 to C_4 remain invariant with respect to the transformation.

images"—that is, the parts of R that will generate I and J from the action of the mapping—are in fact two vertical bands K and L within the original square.

The invariant set of points Λ of the mapping and its inverse T^{-1}, are obviously contained in both rectangle pairs I, J and K, L. They thus are a subset of the four small squares C_1 to C_4 in Figure 56. We call these *recurrent sets*.

Figure 57 shows the result of a second iteration, $T^2 = T \cdot T$, of the map on R. In particular, it displays the bands of $T^2 \cdot R$ that are contained in the original square and their preimages, which are part of the set generated by the action of the inverse mapping T^{-2} on R. We obtain a recurrent set of 16 squares comprising the four solid squares within each of the squares C_1 to C_4 of Figure 56.

By continuing the iterations we see that the points that will never leave R are situated in 4^n squares whose dimensions decrease exponentially to zero as n tends to infinity. The resulting set Λ of recurrent points is composed of a nondenumerable infinity of points, and its total area is zero. It has therefore the features of a Cantor set or, more precisely, of the product of two Cantor sets.

A detailed analysis of the dynamics on the set Λ can be carried out by the technique of *symbolic dynamics*, which will be touched on in later discussion. It can be shown that any vicinity of any point of Λ contains periodic orbits in the

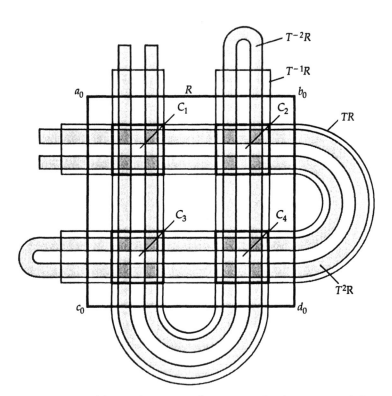

Figure 57 Twice-iterated horseshoe map, illustrating the formation of the set Λ of recurrent points from successive fragmentation of the squares C_1 to C_4.

sense of Section 3.7. Moreover, there exist nonperiodic orbits coming as close as desired to any point of $\wedge$. Now, each point $\wedge$ is by construction a hyperbolic point, since it is at the intersection of a contracting direction (the vertical) and an expanding direction (the horizontal). Thus, two orbits starting initially at two neighboring points of $\wedge$ will diverge exponentially in the course of successive iterations. This implies an extreme sensitivity to initial conditions, one of the main features of the irregular trajectories encountered in the previous section in connection with nonintegrable hamiltonian systems, and of the strange attractors discussed in the next section.

Actually the above analogy goes further. It can be shown that when the separatrices of the unstable points on a Poincaré map bend and intersect as in Figure 55, the original motion generates locally on the Poincaré surface of section a dynamics similar to a horseshoe. This substantiates the use of this model as an architype of irregular motion and particularly of chaotic behavior.

3.10 DISSIPATIVE SYSTEMS IN MULTIDIMENSIONAL PHASE SPACES. CHAOS AND STRANGE ATTRACTORS

We now turn to the implications that the possibilities reviewed in the preceding three sections may have for the behavior of dissipative systems. We have seen that in conservative systems regular and chaotic trajectories can coexist in phase space. Their relative abundance is, however, difficult to estimate, especially when the perturbation from the integrable system is no longer small. In dissipative systems, on the other hand, the property of asymptotic stability implies the possibility of *attracting chaos*. In a system presenting this property all trajectories emanating from a certain well-defined, finite part of phase space would tend sooner or later to the attractor; in other words, for such a system chaos would constitute the rule and would certainly show up in the observed behavior. The main question to be asked in connection with chaotic behavior in dissipative systems is therefore whether these systems indeed give rise to chaotic attractors as some of the control parameters are varied.

The appearance of nonperiodic, turbulentlike behavior in many of the experiments reviewed in Chapter 1 suggests that the answer to this question is yes. It is therefore important to set up models interpreting the observations and to define minimal conditions that must be satisfied for the onset of chaotic behavior. The ideas prevailing until the 1960s suggested a rather discouraging picture for the emergence of turbulence, at least in fluids: turbulence was thought to be the manifestation of the dynamics of an *infinity* of coupled modes, each associated with a particular frequency. Moreover, in order for these modes to become fully coupled, an infinity of transitions occurring beyond the loss of stability of the regular steady-state (laminar) flow appeared to be necessary. It was therefore with great surprise and excitement that scientists subsequently became aware of the possibility that dynamical systems involving only a few variables could present chaotic behavior after a few bifurcations from the reference state.

The first such example was discovered in 1963 by the American meteorologist Edward Lorenz. His model provides a simplified description of the thermal convection problem discussed in Section 1.3, in which only three modes, each associated with a particular spatial wavelength, are retained. The model gives rise to an attractor that captures the principal features of turbulent convection. Such a phenomenon occurs routinely in the earth's atmosphere, and is at the origin of the well-known difficulties in weather prediction. What is more, it suggests that weather and climate are fundamentally unpredictable, since they should share one of the basic properties of chaotic dynamics, which is to depend in a sensitive way on the initial conditions. To stress this most important point it is sometimes said—and it is hardly an exaggeration— that the beating of the wings of a fly in Cambridge, Massachusetts, could well be at the origin of a major climatic change in the Indian subcontinent!

Lorenz's work remained unnoticed until the 1970s, but since then there has been an explosion of new results on chaotic dynamics. The ideas grew out of many disciplines simultaneously: pure mathematics, extensive computer simulations, theoretical and experimental physics and chemistry, and population biology, among others. They have led to interesting prototype equations that give rise to chaotic behavior, as well as to a number of scenarios on how the chaotic attractor is attained as the parameters vary. Let us look at a few selective illustrations of each of these aspects.

Some prototypes giving rise to chaotic behavior

A very interesting mathematical model of chaotic behavior has been suggested by Otto Rössler. It contains three variables and only one quadratic nonlinearity:

$$\frac{dx}{dt} = -y - z$$

$$\frac{dy}{dt} = x + ay \qquad (3.30)$$

$$\frac{dz}{dt} = bx - cz + xz$$

a, b, c being positive constants. The equations of evolution have two fixed points, one of which is the trivial solution $x_s = y_s = z_s = 0$, and the other the point $x_s = c - ab$, $y_s = b - (c/a)$, $z_s = (c/a) - b$. We will discuss only the phenomena occurring around the first fixed point, P_0.

For a large range of parameter values the linear stability analysis reviewed in Section 3.6 and Appendix 1 predicts that P_0 is unstable. The behavior in its vicinity has the following peculiar features. The trajectories are repelled away from P_0 along a two-dimensional surface of phase space, in which the fixed point behaves like an unstable focus; and they are attracted along a one-dimensional curve in a fashion similar to Figure 61. We call such a fixed point a *saddle-focus*. Such a configuration gives rise to instability of motion, a basic ingredient of chaotic behavior, but it also allows for the *reinjection* of the unstable trajectories in the vicinity of P_0 and thus for the eventual formation of a stable attractor.

Figures 58(a)(b) depict the chaotic attractors attained for two different parameter values. Both of the features just mentioned are clearly present. However, in

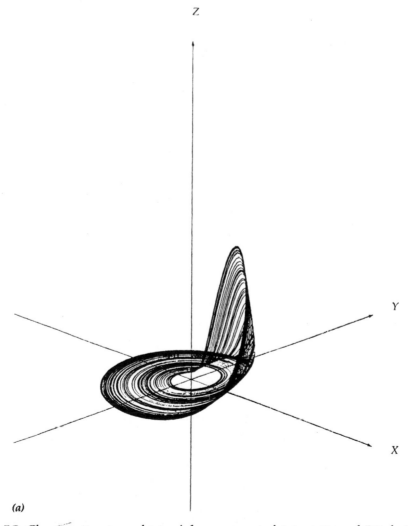

(a)

Figure 58 Chaotic attractors obtained from numerical integration of Rössler's model, Eqs. (3.30). (a) Spiral chaos for $a = 0.32$, $b = 0.3$, $c = 4.5$. The trajectories are injected on the same side of the unstable fixed point. (b) Screw chaos for $a = 0.38$, $b = 0.3$, $c = 4.5$. The trajectories are injected on both sides of the unstable fixed point.

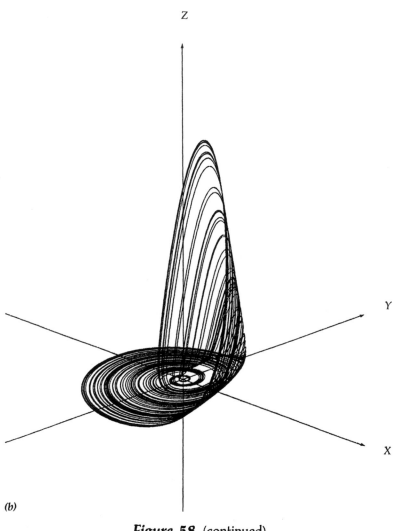

Figure 58 (continued)

Figure 58(a) all the trajectories are reinjected toward P_0 on the same side of the plane xz, whereas in Figure 58(b) some are reinjected on the other side. We call these two motions *spiral* chaos and *screw* chaos, respectively.

Another important feature shown clearly in both of these figures is the *folding* of the surface along which the unstable motion occurs. It is instructive to visualize it on a Poincaré surface of section obtained by cutting the flow transversely with the plane $y = 0$, $x < 0$, $z < 1$. This is illustrated in Figure 59. In Figure 59(a) it is seen that the divergence of the trajectories along the unstable direction $-x$ and their subsequent folding gives rise to a horseshoe map on the Poincaré plane.

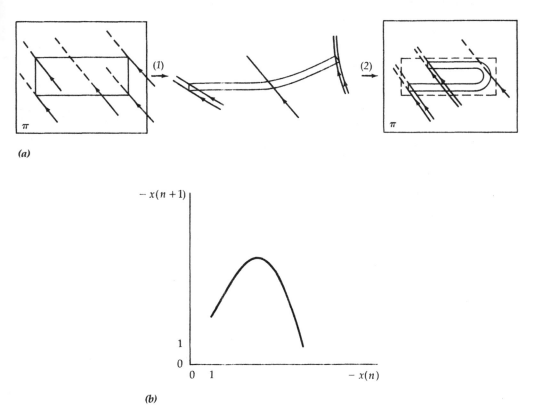

Figure 59 (a) Horseshoe map induced by the dynamical system of Eqs. (3.30) on a Poincaré surface of section. The flow depicted in Figure 58(a) is cut by the plane ($y = 0$, $x < 0, z < 1$). The trajectories emanating from a small rectangle in this plane rotate around the one-dimensional stable manifold of the fixed point and intersect the Poincaré surface repeatedly after having followed the folding of the unstable manifold. (b) One-dimensional map obtained by plotting the value of $-x$ at the $(n + 1)$th intersection of the trajectory with the Poincaré surface, versus its value at the nth intersection.

In view of the results of Section 3.9, this explains qualitatively the random, non-periodic character of the trajectories on the attractor.

A second view of this phenomenon is given in Figure 59(b). Here we plot the position of the $(n + 1)$th intersection point between the above-defined Poincaré surface and the flow, as a function of its position at the nth intersection. The numerical construction shows that we obtain a smooth bell-shaped curve. Note that the positions of the successive intersections on this curve are not given by consecutive points but by points that appear to be distributed randomly.

Figure 59(b) establishes quite naturally the connection between our continuous time model, Eq. (3.30), and a second very important class of dynamical systems showing chaotic behavior. These are discrete time models defined by the iterative equation

$$x_{n+1} = f(x_n) \tag{3.31a}$$

A typical bell-shaped curve similar to Figure 59(b) is obtained by choosing

$$f = 1 - \lambda x^2 \tag{3.31b}$$

In addition to their connection with continuous-time systems through the Poincaré map, systems of this kind can also directly model some situations in biology, where we deal with populations in which the age groups of the individuals present are determined by the age groups of former generations.

Despite their apparent simplicity, Eqs. (3.31) show an incredibly rich behavior ranging from simple fixed points to multiple periodic or chaotic solutions. Moreover, most of the information on the scenarios for reaching chaotic behavior comes from these equations, as we will discuss next. There exists a wealth of experimental examples producing chaotic attractors and giving rise, through suitable Poincaré surfaces of section, to a discrete time dynamics similar to Figure

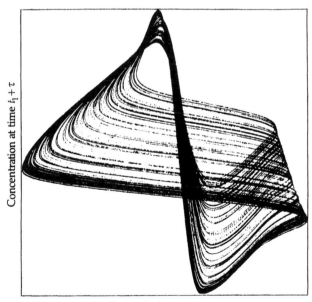

Concentration at time t_1

Figure 60 Chaotic attractor of the BZ reaction in an open system. The signal measured is proportional to the logarithm of the Br^- ion concentration. The coordinates are the signal at various times t, and the signal at times $t + \tau$. The value of time lag chosen in drawing the figure is $\tau = 53$ sec. [From J.C. Roux, R. Simoyi, and H. Swinney, *Physica*, **8D**, 257 (1983)]

59(b). Figure 60 gives a particularly striking illustration obtained from experiments on the Belousov-Zhabotinski reaction in a well-stirred reactor.

Some scenarios for reaching chaotic behavior

In contrast to the static and limit cycle bifurcations discussed in Sections 3.4 and 3.5, the mechanisms of emergence of chaotic attractors cannot be cast in a universal, normal form from which the effect of the parameters can be clearly seen. Rather, by combining some general considerations with the experience acquired from extensive numerical computations, we arrive at regularities that allow us to identify certain particular routes to chaos. It is difficult to assert whether this is an inevitability due to the very nature of chaos, or a temporary drawback due to the complexity of the problem. The situation is very much like that prevailing in the life sciences before the advent of biochemistry and molecular biology, where taxonomy, the art of classifying into groups or families, was the predominant mode of approach.

An important result defining one particular scenario leading to chaos is due to David Ruelle and Floris Takens. As we have seen, one of the attractors that become possible in phase spaces of three or more dimensions is a torus describing a quasi-periodic motion. Such tori arise frequently through secondary or higher bifurcations from periodic orbits. Ruelle and Takens showed that a two-dimensional torus is robust, in the sense of structural stability defined in Section 3.4; in other words, by applying a small perturbation by means of a parameter change such an attractor will generally persist. This is not so for tori of three or more dimensions involving at least three incommensurate frequencies. These are fragile objects and, by disappearing upon the action of a small perturbation, they give rise to a strange attractor that shows the features discussed in Section 3.7 and is structurally stable. Such transitions have been observed in model systems and in experiments, notably in fluid dynamics.

A second scenario that leads to more quantitative predictions arises from the study of the discrete time mappings defined in Eqs. (3.31). It corresponds to an infinite sequence of bifurcations at well-defined parameter values $\lambda_1 < \lambda_2 < \cdots < \lambda_n < \cdots$ each leading to higher order cycles whose periods double at each consecutive bifurcation. The λ_n's accumulate at a particular value λ_∞, after which we obtain "infinite period" orbits showing a markedly chaotic behavior. Eventually the entire state space of this discrete dynamical system (except for a set of measure zero) belongs to a single chaotic attractor displaying instability of motion and sensitivity to the initial conditions.

A most important result, due to Mitchell Feigenbaum, is the existence of regularities in the approach to the chaotic region. It establishes that λ_∞ is attained in geometric progression with the ratio $\delta \simeq 4.669$. Moreover, this behavior depends only on the structure of the iteration function $f(x)$ around its maximum. It therefore is structurally stable and occurs with exactly the same *quantitative* characteristics for whole families of maps, including the simple parabola of Eq. (3.31b), but also a sine wave, an ellipse, and other paths. This is referred to as *metric universality*. Table 3 illustrates the situation. The reader is strongly encouraged to do some "experimental mathematics" and check the table by putting Eqs. (3.31) into a pocket scientific calculator and performing the successive iterations.

Many of the experiments reviewed in Chapter 1 reveal the existence of period-doubling cascades that lead eventually to chaotic behavior, and confirm the theoretical value of the accumulation ratio δ. There is no doubt that the discovery of this trend is a major breakthrough in the field of nonlinear dynamics.

There is a still different scenario for chaotic behavior in discrete time mappings, one known as the Pomeau-Manneville intermittency. It arises when a periodic solution becomes unstable and gives rise to trajectories in which time intervals of practically periodic behavior are followed by intervals of wild behavior in the form of turbulent bursts whose duration is itself random. This pattern has also been observed in most of the experiments surveyed in Chapter 1.

Coming back to continuous time flows, the reinjection of the departing trajectories in the vicinity of an unstable fixed point of the saddle-focus type (Figure 58) is frequently associated with the emergence of orbits of a rather exceptional type known as *homoclinic orbits*. These are trajectories that leave the fixed point but come back to it; in other words, they tend to the same limit when time t goes to $+\infty$ as well as to $-\infty$. Such a trajectory is shown in Figure 61. Homoclinic orbits are very sensitive to variations of parameter values and are generally destroyed if the parameters do not satisfy a strict equality (in the terminology of Section 3.4, they are structurally unstable). However, for nearby values of the parameters their disappearance leaves a very rich structure of orbits in phase space, some of which behave chaotically. There is evidence that some of the above-mentioned roots of chaos involve the appearance of homoclinic orbits. The homo-

Table 3 Period-doubling cascade in the map $x_{n+1} = 1 - \lambda x_n^2$

n (Period of orbit)	Bifurcation point	$\lambda_n - \lambda_{n-1}$	$\delta = \dfrac{\lambda_n - \lambda_{n-1}}{\lambda_{n+1} - \lambda_n}$
0	0.75	—	—
1	1.25	0.5	4.233738275
2	1.3680989394	0.1180989394	4.551506949
3	1.3940461566	0.0259472172	4.645807493
4	1.3996312384	0.0055850823	4.663938185
5	1.4008287424	0.0011975035	4.668103672
⋮	⋮	⋮	⋮
∞	1.4011552⋯	—	4.669⋯
(Nonperiodic behavior)			

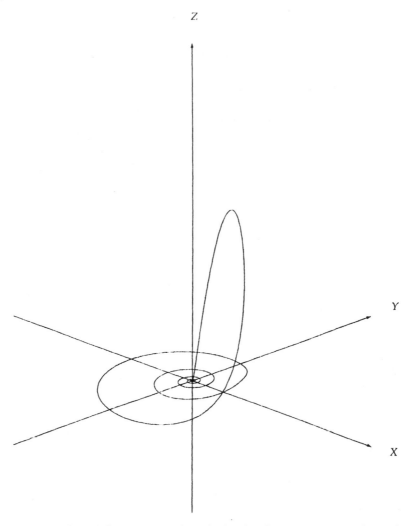

Figure 61 Homoclinic orbit associated with the fixed point $(0, 0, 0)$ of Rössler's model obtained for parameter values $a = 0.38$, $b = 0.30$, $c \simeq 4.82$.

clinic scenario is thus likely to play an increasingly important role in chaotic dynamics.

There are two more issues regarding chaotic behavior.

First, can any statement be made on the relative importance of chaotic versus regular behavior in dissipative systems involving many variables? The experience accumulated from the study of model and real-world systems suggests that both types of behavior are possible and can even coexist. There is therefore nothing like an irresistible tendency of multivariate systems to burst into chaos immediately after the first bifurcation. What is more, for certain types of kinetics it turns out that locating the chaotic domain in parameter space is a very subtle enterprise. Biochemical regulatory processes provide a characteristic example, perhaps not fortuitously so since reproducibility must be an essential feature of biological rhythms. In this respect it is useful to reiterate some earlier comments concerning what is to be regarded as generic or typical. The behavior of physical systems is not necessarily reflected by randomly chosen abstract models. This is especially obvious in biology, in which each function present in living cells is the result of a long evolution. We have here another illustration of the specificity of nonequilibrium systems as opposed to equilibrium systems. Indeed, if we lower the temperature of any material under equilibrium conditions, eventually we will reach a phase transition point. But if we control a parameter interfering with the behavior of a nonequilibrium system, we can observe limit cycles, steady states similar to equilibrium, chaos, or even something different. We can evolve from thermal, equilibriumlike disorder to large-scale order and then reach this peculiar state of small-scale disorder and large-scale order that is chaos; or we can witness an opposite evolution or even miss chaos altogether.

The second issue concerns the role of chaos in nature. We have stressed the fact that chaotic attractors provide the interpretation of several experimental results. Beyond this successful confrontation, however, chaos reveals to us a whole new world of forms and patterns. It shows that disorder in a certain range is perfectly compatible with order at a different range, as is revealed by the very existence of an attractor. It also shows that randomness is not the result of experimental imperfections or of a complex environment that we cannot control, but rather is deeply rooted in the dynamics of perfectly deterministic systems involving few variables. Above all, the coexistence of both randomness and order gives rise to a new concept that has so far been absent in our discussion, namely, *information*. We will analyze this newcomer to the world of complexity briefly in Section 3.12 and more extensively in Chapter 4.

3.11 SPATIALLY DISTRIBUTED SYSTEMS. SYMMETRY-BREAKING BIFURCATIONS AND MORPHOGENESIS

So far in this chapter we have been dealing with dynamical systems involving a finite number of variables. We now turn to spatially distributed systems. Here the state variables (say the temperature in a slab subject to a systematic heating in one of its sides) are fields, in the sense that they depend continuously on the space coordinates. This implies that there is an infinity of variables present in the problem. Generally speaking, these variables will be coupled, since the spatial inhomogeneities will give rise to transport phenomena such as heat conduction, diffusion, or convection. This will be reflected by the presence of *spatial derivatives* in the evolution equations, like the laplacian of temperature $\nabla^2 T(r, t)$, or concentration $\nabla^2 c(r, t)$, the gradient of the velocity $\nabla v(r, t)$, and so on. We expect that such systems should be much more difficult to handle; however, in many instances interesting results can be obtained by using suitable perturbative procedures that reduce the problem to a finite number of variables.

This mechanism is quite similar to that outlined in Section 3.6 for the reduction of the dynamics of a multivariate system to a normal form. The starting point is again linear stability analysis (see Appendix 1). In the presence of spatial inhomogeneities, the linear stability operator L appearing in Eq. (3.25) will contain spatial derivatives. But in many cases these derivatives occur in an additive fashion [see, for instance, Eqs. (A1.14)]. We then look for solutions that are the product of a time-dependent part, $c(t)$, and a function $u(r)$ containing all space dependencies [Eq. (A1.19)]. This leads to a separation of variables, that is, to two simpler sets of equations involving only $u(r)$ and $c(t)$. The equations for $u(r)$ can frequently be solved exactly. For instance, in a one-dimensional system involving diffusion, $u(r)$ is a superposition of sine and cosine functions—the eigenfunctions of the laplacian in the spatial domain of interest. The time-dependent part $c(t)$ can then be handled exactly as in Eq. (3.25), and a graph similar to Figure 48 describing the rate of growth of the perturbations versus the control parameter λ can be deduced. However, a new element comes into play: Some of the coefficients $L_{ij}(\lambda)$ contain a "remnant" of the spatial inhomogeneities, namely, the action of the space derivatives on the function $u(r)$ or equivalently the eigenvalue of the space-dependent part of the operator L. The latter depends on the value of the transport coefficients and on the characteristic spatial length of the mode described by $u(r)$. In a reaction-diffusion system this is merely given by the products $-D_i k^2$, where D_i is the diffusion coefficient of species i and $k = 2\pi/l$, l being the characteristic space length [see Eq. (A1.20)].

Having this additional freedom, we can ask: Suppose that only one of the γ versus λ curves in Figure 48 crosses the λ axis. At the marginal state $\lambda = \lambda_c$, the characteristics of the system will still depend on the additional parameter k. Is it possible to identify the most "dangerous" value of k that is likely to dominate the transition to instability? To answer this question it is instructive to plot the critical parameter value λ_c versus k. Two typical results, corresponding respectively to Eqs. (A1.23a) and (A1.23b), are described in Figure 62.

In both cases a and b shown in the figure, the λ_c versus k curve has an extremum, λ_c' and λ_c'', respectively. This means that as λ is varied from values corresponding to asymptotic stability (below curves a and b) to values leading to instability (above curves a and b), the *first* transition will take place at λ_c' or λ_c'', and will therefore dominate the behavior of the system for nearby values of λ. Now in case a, the extremum λ_c' occurs at $k = 0$, that is, at a characteristic spatial length $l = \infty$. This obviously corresponds to a space-independent situation. In other words, the dominant mode in the vicinity of the first bifurcation point will be a homogeneous one. The analysis will thus reduce to that carried out in the previous sections.

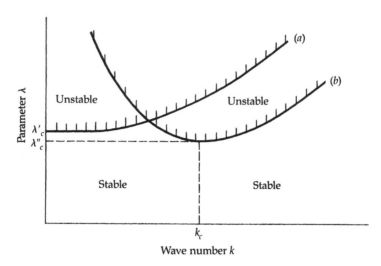

Figure 62   Two typical outcomes of linear stability around the uniform steady state of a spatially distributed system. Marginal stability curve (*a*) has an extremum at wave number value $k = 0$. On these grounds one expects that the dominant mode beyond the instability will remain spatially homogeneous. Marginal stability curve (*b*) has an extremum at $k = k_c \neq 0$. The bifurcating branches emerging beyond the transition are expected to display a nontrivial space dependence. This is the basis of the phenomenon of space symmetry breaking.

But a new and exciting possibility arises in case *b* of Figure 62. Here the extremum λ_c'' occurs at a nontrivial value $k = k_c$, that is, at a well-defined characteristic spatial length $l = l_c$. It follows that the dominant mode in the vicinity of the first bifurcation will now be spatially inhomogeneous. Moreover, its characteristics will be *intrinsic*, in the sense that they will be determined entirely by the system's parameters. We call this situation *space symmetry breaking*.

In both cases *a* and *b* we can again define a suitable set of order parameters that obey a closed set of equations in the vicinity of the first bifurcation. The *normal form* of these equations has the same structure as in Sections 3.4 to 3.6 and Appendix 2. Moreover, when several modes become unstable simultaneously, we can extend the description in the form of a larger number of equations describing the interaction between several order parameters. This gives rise to secondary and higher order bifurcations leading to, among other phenomena, the coexistence of several stable spatial patterns. Finally, in systems of large spatial extension the number of interacting modes tends to infinity and as a result, the normal form equations for the order parameters become partial differential equations. Still, one may show that even for such systems the trajectories tend to a finite dimensional attractor embedded in a finite dimensional phase space, referred to as the *inertial manifold*. These dimensionalities may become quite high if the system operates far from the instability threshold.

The onset of spatial structure in a hitherto homogeneous medium is a central problem in many fields. In hydrodynamics, space patterns frequently accompany the transition to instability, as in the Bénard problem described in Chapter 1. In interfaces and materials science space patterns give rise to intriguing forms of which dendrites are a striking example. In embryonic development space structures provide the necessary positional information for cell differentiation, as discussed briefly in Section 2.7. In this latter case—and indeed in all problems involving a diffusionlike coupling between variables—an interesting feature is the existence of a *minimal system size* below which spatial patterns cannot form. This can be understood easily on the intuitive level. In a spatially distributed reactive system, chemical reactions and diffusion cooperate in a symmetry-breaking bifurcation in the following fashion. Chemical kinetics, through its intrinsic feedbacks, may give rise locally to a runaway phenomenon by amplifying the effect of small perturbations or fluctuations. Diffusion tends to smear out the inhomogeneities caused by this, but when its rate is comparable to the reaction rate, it does not quite succeed in erasing them. The result is then a space pattern whose characteristic length reflects the average distance over which a group of reactive molecules can diffuse before a reaction takes place. Now, in a small-size system reactions occur at the same rate as in a large system, but diffusion is tremendously enhanced (the diffusion rate is proportional to D/l^2, D being the diffusion coefficient and l the system size). The system becomes dominated by the "stirring" imposed by diffusion and the boundary conditions wipe out any local attempt at a destabilization.

Space patterns arising through symmetry-breaking bifurcations can be stationary as well as time-dependent. An example of stationary patterns is given in Figure 63. A set of two coupled reaction-diffusion equations has been simulated

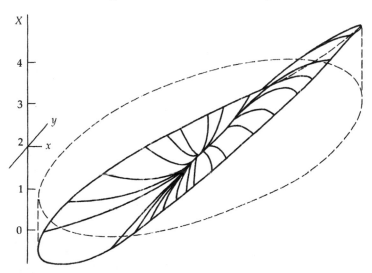

Figure 63 Perspective view of a spatial concentration pattern X generated on a circular domain (in x–y plane). The pattern comes from the product of a radial part, the first-order Bessel function $J_1(r)$, and an angular part, cos φ.

numerically on a circular disc. We see clearly the appearance of a *polarity axis* in the system, inducing a differentiation along a preferred direction which in a real-world system might be, for instance, an up-down or a dorso-ventral axis.

Striking experimental examples of time-dependent space patterns are the waves observed in the BZ reagent and in *Dictyostelium discoideum*, shown in Figures 9 and 15. Such patterns can also be obtained from model systems. Two examples giving rise to simple rotating waves and rotating spirals are given in Figure 64.

One characteristic property of both stationary and time-dependent space patterns, already mentioned in Section 2.7, is to appear in the form of *multiplets*, each

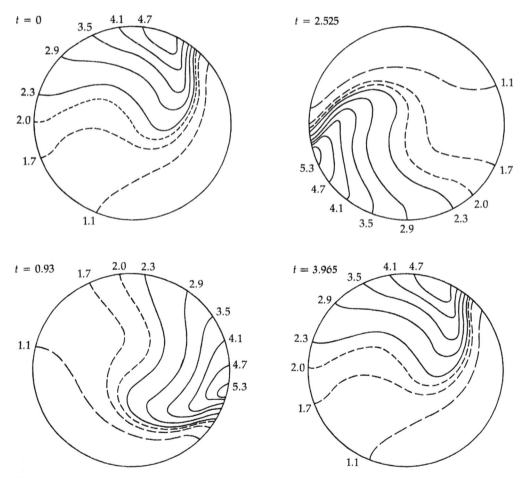

Figure 64(a) Rotating waves obtained by numerical solution of the brusselator model, Eqs. (3.19), in a circle of radius 0.5861 subject to zero flux boundary conditions. Solid and broken lines refer, respectively, to constant concentrations larger or smaller than values on the homogeneous (unstable) steady state. Parameter values: $A = 2$, $B = 5.8$; diffusion coefficients of species X, $D_X = 0.008$, and Y, $D_Y = 0.004$.

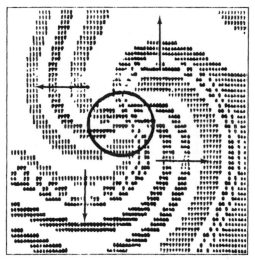

$t = t_0$

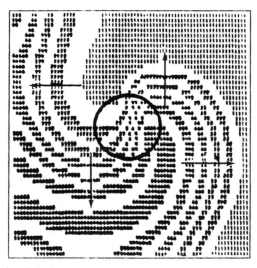

$t = t_1 > t_0$

Figure 64(b) Rotating spirals obtained by numerical solution of model

$$\partial A / \partial t = \nabla^2 A - A - B + (A \text{ if } A > 1/20)$$
$$\partial B / \partial t = \nabla^2 B + A/2$$

[from A. Winfree, *SIAM-AMS Proceedings*, **8**, 13 (1974)]

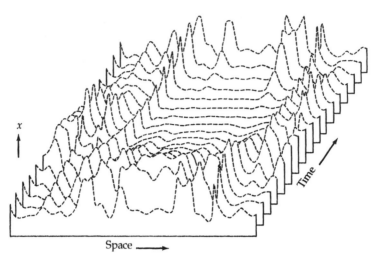

x

Time

Space ⟶

Figure 65 Representation of diffusion-induced chaos obtained from numerical solution of the brusselator model. [From Y. Kuramoto, *Physica*, **106A**, 128 (1981)]

member of which has identical structure and stability properties in a symmetric medium. For instance, in Figure 63 the polarity axis could be any diameter of the circle. Similarly, in an infinitely extended oscillating medium spirals appear in pairs, the two members of which rotate in opposite directions (clockwise and counter-clockwise). Thus, each spiral is an asymmetric pattern displaying a preferred *chirality*, but in a population of spirals in a macroscopic system symmetry is restored by the coexistence of equal numbers of forms of opposite handedness. The problem of selection between such forms is discussed in Section 3.13.

An intriguing class of space and time-dependent patterns arises when diffusion or some other transport phenomenon induces a chaotic dynamics in an otherwise regularly behaving homogeneous system. Figure 65 depicts a typical configuration obtained by numerical simulations on the brusselator model. Similar behavior arises in coupled nonlinear oscillators where, depending on the strength of the coupling, the system can be driven from a regime of perfect phase locking to a chaotic regime characterized by marked phase desynchronization.

3.12 DISCRETE DYNAMICAL SYSTEMS. CELLULAR AUTOMATA

In many instances a physical phenomenon involves a sharp threshold. Beyond such a threshold the rate of a process levels off to some value that is quite different from the value prevailing below the threshold, and quite often the interval of transition between these plateaus is very small. For instance, an electronic circuit involving an active element like a transistor will switch between two current levels over a very short interval of voltage values. Such circuits are the building blocks of computers. In biology, a gene directs the synthesis of a product, say a protein, in such a way that the concentration of the latter is practically zero when the gene is inactive, a fact that may in turn be influenced by the concentration level of an effector coupled to the protein activity, and so forth. In this section we briefly describe the dynamical behavior of systems whose elements present threshold phenomena.

Consider first a simple element involving only one variable, X. The following is a typical situation: for $0 < X < X_0$ the rate of change $F(X)$ is very small, whereas for $X > X_0$ it suddenly becomes appreciable and saturates shortly thereafter to some value F_{max}; this is shown in Figure 66. To capture the essence of this response we regard F as a discontinuous function of the discontinuous variable X. In other words, we consider that in $0 < X < X_0$ both X and F are "zero," whereas for $X > X_0$, X and F are equal to "one." Such quantities are known as *boolean variables*, after the nineteenth-century English mathematician George Boole, who is regarded as the founder of formal logic.

The time evolution of X will be dictated by appropriate logical equations linking X and F and which are the substitutes of the differential equations written in the traditional continuous formalism [see, e.g., Eq. (3.1)]. The simplest example is provided by

$$X_{t+1} = F(X_t) \qquad (3.32)$$

in which the variables present are changing values in full synchrony and in discrete time steps. The form of F depends on the specific problem considered. For instance, let a positive feedback loop be involved (an example is the ice-albedo

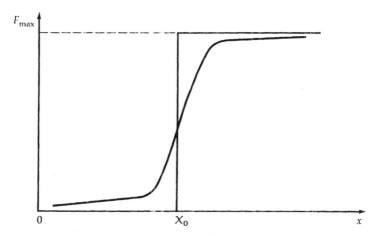

Figure 66 Representation of a boolean function F, of a boolean variable X.

feedback discussed in Section 1.8). In logical terms this means that if X (the temperature in the albedo example) is high the dynamics will tend to increase it further; in other words, the value of $F(X)$ will also be high, and vice versa if X is low. Remembering that F and X are boolean variables, we may express this dependence by the logical equation $F = T$.* The solutions of this equation are nicely visualized in a state table in which the logical values of F and X are successively plotted:

F	X
0	0
1	1

We see that there exist two states, one (0, 0) in which F and X are simultaneously off, and one (1, 1) in which they are simultaneously on. Such states can no longer evolve in time and are the analogues of fixed points of continuous dynamical systems. We have here a very simple model of the ability of positive feedback systems to undergo transitions between different steady states. More complex situations involving many variables, as well as both positive and negative feedback, can be analyzed straightforwardly.

A very interesting situation arises in systems consisting of a lattice of identical boolean components interacting locally in space. Such dynamical systems are referred to as *cellular automata*. A typical one-dimensional cellular automaton is described by

$$X_{i,\,t+1} = F_i(X_{i-r,\,t},\, X_{i-r+1,\,t},\, \ldots,\, X_{i,\,t},\, \ldots,\, X_{i+r,\,t}) \qquad (3.33)$$

which states that the (boolean) variable X of a particular site i is determined by the previous values of a neighborhood of $2r + 1$ sites around it. Depending on the rules involved in F_i, dynamical systems of this type and their two-dimensional analogues can generate a variety of quite spectacular spatial patterns very similar to those described in Section 3.11.

We now ask a different kind of question. An array of "processors" operates on (boolean) data received locally from the neighbors, as for example in Figure 67. Each element, whose internal state is also represented by a boolean function, "computes" its output, which is then transmitted to its neighbors. Could a network of this kind give rise to global "computation" patterns as a function of time featuring some new, *emerging properties* not included in the initial "programmed" units? In principle, the answer to this question is yes. It is based on the existence of attractors of the underlying (dissipative) dynamical system. An attractor corresponding to a self-organizing pattern emerging through a symmetry-breaking instability necessarily endows the system with a collective property that tran-

---

* Notice that despite its apparently linear form this equation actually expresses the highly nonlinear dependence depicted in Figure 66.

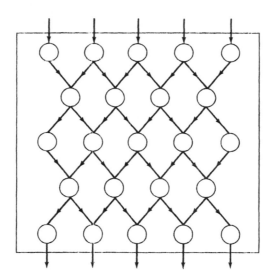

Figure 67 An array of processors, each of which operates on data received locally from its neighbors.

scends the properties of the individual subunits. Actually, in a system of large spatial extension there will be many attractors. Depending on the initial stimulus, the system will end up in different basins of attraction. This allows one to foresee the possibility of machines learning reliably how to recognize different inputs, even when the inputs are slightly distorted.

The role of attractors in computation—or in thinking, for that matter—raises a number of fascinating issues at the interface between dynamical systems theory and nonequilibrium physics on the one side, and computer and cognitive sciences on the other side. The results that will arise from this unexpected synthesis are likely to be far-reaching indeed.

3.13 ASYMMETRY, SELECTION, AND INFORMATION

We have seen that spatially asymmetric solutions arising through bifurcation in an initially perfectly symmetric medium have a special status. In a particular physical situation only one such solution is realized in a sufficiently small region of space. For instance, as we saw in Figure 9, a particular part of the BZ reagent becomes a center from which only a spiral of given chirality, say left-handed, can emanate. In this sense we may speak of *symmetry breaking*. However, if we look at the same part of the system in a large number of uncorrelated experiments, we will observe that solutions of opposite symmetries occur with equal frequency. Similarly, in a given experiment carried out in a system of large spatial extension, solutions of opposite symmetries will coexist in different parts of the system. In this sense, therefore, the symmetry of the initial medium will be restored.

In usual laboratory conditions the above conclusion was to be expected. We cannot see why a Bénard cell rotating, say, clockwise rather than counterclockwise, should appear preferentially in a particular place of the apparatus. Similarly, in the BZ reagent experiments have always shown that spiral waves appear by pairs unless special care is taken by the experimenter to favor a particular configuration.

In many other instances, the inability to select one particular asymmetric state of matter among the multitude of forms available leads to some very basic questions. In order to understand the nature of the problem, let us perform the following thought experiment, summarized in Figure 68. Suppose that in an appropriate chemical reagent a certain state of rotating spiral waves arises. Let $(P_1, Q_1, \ldots)(P_2, Q_2, \ldots)$ be centers of right- and left-handed spirals, respectively, in this specific state. Suppose further that by suitable circuitry a left-handed spiral is made to generate the symbol "D" on a screen, whereas by similar circuitry a right-handed spiral generates the symbol "O." One possible use of these symbols is as follows: When the screen shows the signal released by P_2 above the signal released by P_1 reading from top to bottom the resulting pattern is an instruction to perform a specific task: "DO." It is of course assumed that the text constructed by this algorithm can be read and understood by the reader who will be assigned to perform the task. In the example chosen, this will be so if the reader has a rudimentary knowledge of English.

Let us follow the functioning of this machinery as the reagent is brought to the regime of rotating wave solutions. In the first experiment we choose P_1 and

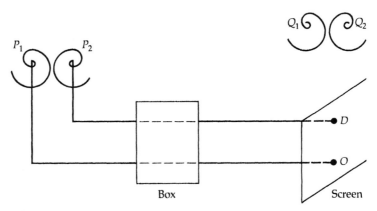

Figure 68 A "bifurcation machine." By endowing a system with a multiplicity of solutions, bifurcation potentially creates information. Symbol D or O is released depending on whether a spiral wave is left- or right-handed. Without selection the handedness of the particular wave that will occur in successive operations cannot be reproduced. That is, there is no way to assure an identical pair Q_1, Q_2 will follow P_1, P_2. In that case the succession of the symbols D and O will occur at random, and this will compromise the processing of the information.

P_2 in such a way that P_2 was the center of a left-handed spiral and P_1 the center of a right-handed spiral. Thus the instruction "DO" appeared on the screen, was properly interpreted and the assigned task was performed. After some time of successful functioning, the machine was switched off. Now let us see what may happen when we use it once again.

Because of the multiplicity of regimes available, points P_1 and P_2 can independently become centers of either left or right spirals. Thus, our machine will not necessarily generate the same message "DO" as before. Instead, there is a good chance that our reader will see on the screen a message that he cannot comprehend: "OD," "OO," or "DD." As a result he will not proceed to perform the task assigned to him. In other words, the "bifurcation machine" we have devised turns out to be an unreliable *information processor* for the purpose of getting our task completed.

Suppose however that we bias the situation. After the first functioning of the machine, we install two small motors below points P_1 and P_2, respectively stirring the medium slightly to the right and to the left. Now a left-rotating wave will invariably be selected in the vicinity of P_1 and a right-rotating wave in the vicinity of P_2. As a result, the message "DO" will be transmitted each time the system functions. Obviously we will now have a reliable machine.

The motors we introduced into our simple example are an illustration of the general concept of *selection*. In other words, space symmetry breaking is the necessary prerequisite without which the possibility of constructing an information processsor would simply not exist. But it is selection that enables us to detect, interpret, and transmit the "message" hidden in the nonlinear, nonequilibrium dynamics of the systems. *Selection decodes information*, and that allows the transfer of complexity from one level to another.

But the need for selection goes far beyond our simple example. In section 1.9 we mentioned the breaking of symmetry between matter and antimatter. It is because matter was "selected" over antimatter in the present universe that nuclei, galaxies, and living beings could form instead of an on-going matter-antimatter annihilation that produced only photons. Closer to our scale, selection of preferred asymmetric forms is at the basis of the structure and functioning of present-day living organisms. Let us survey a few characteristic illustrations showing the ubiquity of asymmetry in this field.

We know from elementary biology that proteins are among the basic building blocks of life. Independently of their origin and function, all the amazing variety of existing proteins arises from the linkage of the same 20 *amino acid* species in different molar proportions. The amino acids found in proteins have four groups bonded to a central carbon atom, called the α carbon:

$$H_2N \overset{\displaystyle COOH}{\underset{\displaystyle R}{\vdots C \vdots}} H$$

The four groups are:

1. A carboxyl, abbreviated COOH
2. An amino group, NH_2
3. A hydrogen atom, H
4. A side or R group that is different for each species

In all of the amino acids except glycine, for which R is identical to H, the α carbon is *asymmetric* because it has different groups attached to it. There are thus two possible amino acid configurations, the D isomers and the L isomers:

$$\begin{array}{cc}
\text{COOH} & \text{COOH} \\
| & | \\
\text{H---C---NH}_2 \qquad & \text{NH}_2\text{---C---H} \\
| & | \\
R & R \\
\text{D} & \text{L}
\end{array}$$

A pair of D and L isomers cannot be brought to coincide by rotations. They are like a right and left hand and are analogous in this respect to the pair of spirals generated in the BZ reagent. Their asymmetry is reflected by their ability to rotate the plane of polarization of polarized light in a preferred direction, to the right for the D isomers, and to the left for the L isomers.

In proteins, amino acids are always found in the L configuration. We therefore have a striking example of an absolute selection of asymmetry. The great French scientist Louis Pasteur, the founder of modern biochemistry, regarded this phenomenon as one of the basic aspects of life.

But we are far from exhausting the list of examples of asymmetry in biology. Nucleic acids, the other basic building blocks of life along with proteins, are composed of subunits known as nucleotides. In addition to a purine or pyrimidine base—essentially a planar molecule—they contain a sugar displaying a three-dimensional chiral structure: D-deoxyribose in the case of deoxyribonucleic acid (DNA), and D-ribose for ribonucleic acid (RNA). Again, all nucleic acids of living organisms appear to be the result of an absolute selection in which only the right-hand isomer D is found.

A still different, but equally striking, case of preferred asymmetry refers to how the genetic code is read. It is well established that the present code is a triplet code, each symbol of which corresponds to a set of three nucleotides (codons). Most important, there exist *start signals* that allow the genetic message to be "read" in groups of three nucleotides in a *fixed direction* from the starting point. The importance of this asymmetry is primordial, as its absence would produce effects analogous to reading English text indifferently from left to right or right to left. In this respect it is worth noting that synthetic nucleotides, which lack this polarity, have very reduced information capacities.

By now we should be convinced of the constructive role of selection. It is therefore important to inquire about the mechanisms by which selection could come about.

The most clear-cut case of selection arises when the symmetry-breaking bifurcation is a unique, nonrepeatable event. One example is the selection of matter over antimatter which, once performed, leads the universe to a point of no return beyond which it is impossible to imagine a different realization—at least at a time scale of 20 billion years! A similar case arises if the transition to the asymmetric form is not spontaneous, but involves a stage of nucleation in which the system must first become "activated" and overcome a "barrier." If the amount of activation energy required is very large, the transition to a particular asymmetric state is very slow. It is thus conceivable that by the time the state of opposite symmetry was about to arise, the first symmetry would have taken over, modified the environment, and inhibited the subsequent appearance of any form of different symmetry.

A second general mechanism of selection is the transfer of an external asymmetry to the particular system under consideration. This immediately raises two issues. First, this "primordial" asymmetry must always act in the same way. One example in which this condition is satisfied is gravity, which inevitably influences any earth-bound experiment involving objects of macroscale mass by introducing a preferred up-down direction. Second, the system must be sufficiently sensitive to perceive the effects of the external asymmetry, which usually are very weak. It is here that bifurcations again play an important role. As we saw in Section 3.4, bifurcation is deeply affected by small changes of the parameters appearing in the equations of evolution of the system and can even be destroyed under certain conditions [see, e.g., Figure 42(c)]. Therefore, if during its history a system happens to come sufficiently close to a bifurcation point for an extended period of time, it will become sensitive to small effects that conceivably could induce selection of a particular form of asymmetry. We discuss this mechanism further in Chapter 4, in which the concepts necessary to handle stochastic perturbations are developed.

In both the above mechanisms irreversibility is present, ensuring the asymptotic stability of the different states. And we can envisage another, different mechanism of selection in which irreversibility plays an even more direct role. In particular, we may ask whether the universal time-reversal symmetry breaking associated with irreversibility can itself be a source of selection leading to information-rich *spatially* asymmetric forms of matter. At this time we are not in a position to handle this question. We first have to define the concept of information properly, for up to this point we have referred to information rather loosely, using the language of intuition. We can achieve such a definition only after examining the stochastic theory of dynamical systems.

Randomness and Complexity

*P*reviously we alluded to the role of random elements in the emergence of complex behavior. In all physico-chemical systems such elements are continuously present by the mechanism of internal *thermodynamic fluctuations*. Furthermore, in many circumstances the environment impinging on a system is also noisy—for example, an ecosystem that is buffered by seasonal or other climatic fluctuations. In this chapter we deal with the dynamics of the fluctuations. Special emphasis is placed on the mechanisms by which they interfere with the macroscopic behavior, and on the new concepts brought about by a probabilistic approach to dynamical systems. As we will see, stochastic—probabilistic—effects play an especially crucial role in the vicinity of bifurcation points.

4.1 FLUCTUATIONS AND PROBABILISTIC DESCRIPTION

The observation of the state of a physico-chemical system usually involves an averaging of the instantaneous values of the pertinent variables, either over time or over a volume of supermolecular dimensions. For instance, if we put some 0.33×10^{23} molecules of water in a container of one cubic centimeter ($1\ cm^3$) volume at ambient temperature and pressure, we would conclude that we have a liquid whose number density is 0.33×10^{23} molecules/cm^3 and whose mass density, in grams, is $0.33 \times 10^{23} \times$ (mass of H_2O molecule) $= 1\ g/cm^3$. Number and mass densities are the sort of variables we have dealt with in most of the preceding chapters, but there we looked primarily on the macroscopic scale. The number density of a liquid in a small volume element, of the order of, say, ten cubic angstroms ($10\ Å^3$), will continuously deviate from its value in the surrounding macroscopic volume. Molecules will cross the boundaries of microscale volume elements, and because of the random character of their motion the number of particles contained at any moment in each small volume will be essentially unpredictable. We call the deviations generated by this mechanism fluctuations. Because of fluctuations, physico-chemical systems are capable of exploring the phase space continuously and of performing excursions around the state predicted by the solution of the phenomenological, deterministic equations that describe the systems.

The natural approach to the problem of fluctuations is in terms of probability theory. For instance, in a system undergoing chemical reactions the quantities of interest are the probabilities $P(X_\alpha, \Delta V, t)$ of having X_α particles of species α in a volume element ΔV at time t. More complete information will be contained in the multivariate probabilities $P(X_{\alpha j}, \Delta V_j; X_{\beta k}, \Delta V_k; \ldots, t)$ of the simultaneous occupation of various volume elements j, k by the chemical species α, β, $\ldots$

A very convenient way to characterize a probability distribution is to study its *moments*, that is, the set of quantities

$$\langle X_{\alpha j}^{m_{\alpha j}} \cdot X_{\beta k}^{m_{\beta k}} \cdots \rangle = \sum_{X_{\alpha j}, X_{\beta k} \cdots} X_{\alpha j}^{m_{\alpha j}} X_{\beta k}^{m_{\beta k}} \cdots P(X_{\alpha j}, \Delta V_j; X_{\beta k}, \Delta V_k; \ldots) \quad (4.1)$$

$m_{\alpha j}$, etc. being integers. Two particularly interesting moments are the average value, $\langle X_{\alpha j} \rangle$ and the second moment $\langle X_{\alpha j} X_{\beta k} \rangle$. Actually, in most cases it is preferable to argue in terms of an excess quantity known as *covariance*:

$$\langle \delta X_{\alpha j}\, \delta X_{\beta k} \rangle = \langle X_{\alpha j} X_{\beta k} \rangle - \langle X_{\alpha j} \rangle \langle X_{\beta k} \rangle \quad (4.2)$$

Let us illustrate the significance of these quantities in the particular case of a single stochastic variable, X, for which expression (4.2) reduces to the *variance*, $\langle \delta X^2 \rangle$, of the underlying probability distribution. We first introduce a statistical measure of dispersion called the (normalized) *standard deviation*

$$\delta = (\langle \delta X^2 \rangle / \langle X \rangle^2)^{1/2} \tag{4.3}$$

A small δ implies a distribution that is sharply peaked about its mean value, $\langle X \rangle$, which in turn is close to the most probable value, $\bar{X}$, for which $P(X)$ takes its unique maximum. An extreme example of such a situation is a probability distribution of the form of an infinitely sharp peak $P(X) = \delta^{kr}_{X,\bar{X}}$ that is,

$$
\begin{aligned}
P(X) &= 1 &&\text{if } X = \bar{X} \\
&= 0 &&\text{if } X \neq \bar{X}
\end{aligned}
\tag{4.4}
$$

In other words, all states except a particular one, $X = \bar{X}$, are unoccupied. The mean and variance are obviously given by

$$\langle X \rangle = \sum_X X\, \delta^{kr}_{X,\bar{X}} = \bar{X}$$

$$\langle X^2 \rangle = \sum_X X^2\, \delta^{kr}_{X,\bar{X}} = \bar{X}^2$$

and

$$\langle \delta X^2 \rangle = \langle X^2 \rangle - \langle X \rangle^2 = 0$$

or, alternatively, $\delta = 0$. A more general distribution satisfying the requirement that δ is small is depicted in Figure 69(a). The poissonian distribution, which we encountered in Section 2.8, belongs to this class. Moreover, the explicit calculation of δ performed at the end of Section 2.8 shows that for this distribution $\delta = 1/N^{1/2}$, N being a large number measuring the size of the system (e.g., through the mean number of particles contained in a volume V, or through the value of V itself). Now, in systems of practical interest the linear dimension of V is very large compared to the mean separation between molecules. Formally, this is expressed by the *thermodynamic limit*, $V \to \infty$. We see that in this case δ goes to zero. In other words, there is a clear-cut separation between macroscopic behavior and fluctuations. Many of the other basic distributions encountered in probability theory, for example, binomial or gaussian distribution, share this important property.

The above features can be expressed in an elegant and general form by two basic results of probability theory known as the *law of large numbers* and the *central limit theorem*. In both theorems we are interested in the properties of a variable X that is the sum of statistically independent random variables with a common distribution:

$$X = X_1 + X_2 + \cdots + X_n \tag{4.5}$$

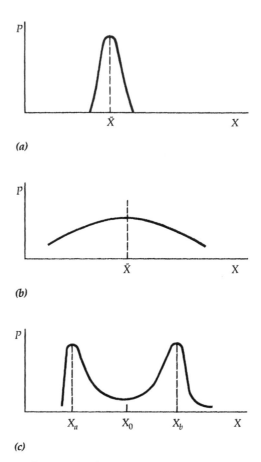

Figure 69 Stochastic analogue of bifurcation. As the system crosses the critical value λ_c of a control parameter λ the probability function changes form. (a) $\lambda < \lambda_c$; a unimodal form peaked sharply on a unique attractor. (b) $\lambda = \lambda_c$; a flat distribution. (c) $\lambda > \lambda_c$; a multihumped distribution whose maxima coincide with the new attractors emerging beyond bifurcation.

Under the very mild assumption that the average value $\langle X_k \rangle \equiv m$ of the individual distributions exists, the law of large numbers asserts that the probability that the arithmetic mean $(X_1 + \cdots + X_n)/n$ differs from the statistical average m by less than any prescribed value ε, tends to one as $n \to \infty$ and $\varepsilon \to 0$. Under the additional assumption that the individual variance $\langle \delta X_n^2 \rangle \equiv \sigma^2$ exists, the central limit theorem asserts that for the limit $n \to \infty$ the probability distribution of $[(X_1 + \cdots + X_n)/n - m]/(\sigma/\sqrt{n})$ becomes as close as desired to a *normal distribution*, the integral of the *gaussian density function*. It follows that the variance of $(X_1 + \cdots + X_n)/n$ is of the order of $\sigma^2 n^{-1}$, whereas the variance of X itself is $\sigma^2 n$. The latter is precisely the estimate we arrived at by the explicit calculation of the poissonian variance, where the role of n is played by the size parameter N, which expresses the variable X as the sum of the observable partial values of small volume elements of the system. Such an X is termed an *extensive variable*.

Results of this kind dominate the literature of probability theory; however, it is very easy to construct counterexamples. Consider, for instance, a probability distribution of a macrovariable (an extensive variable, or an intensive variable constructed by dividing an extensive variable by the system size), of the form shown in Figure 69(c). Clearly this is not a gaussian distribution. It is a two-hump distribution whose two maxima, at $X = X_a$ and $X = X_b$, are separated by a minimum at $X = X_0$. Let us idealize the situation further by assimilating P to two infinitely sharp peaks,

$$P = \tfrac{1}{2}(\delta^{kr}_{X,X_a} + \delta^{kr}_{X,X_b}) \tag{4.6}$$

where the factor $\tfrac{1}{2}$ ensures that $P(X)$ is normalized to one,

$$\sum_{X=0}^{\infty} P(X) = 1$$

We compute the mean value and the variance:

$$\langle X \rangle = \sum_X X\, P(X) = \tfrac{1}{2}(X_a + X_b)$$

$$\langle X^2 \rangle = \sum_X X^2 P(X) = \tfrac{1}{2}(X_a{}^2 + X_b{}^2)$$

and

$$\langle \delta X^2 \rangle = \tfrac{1}{2}(X_a{}^2 + X_b{}^2) - \tfrac{1}{4}(X_a + X_b)^2$$
$$= \tfrac{1}{4}(X_a - X_b)^2$$

This leads to a statistical dispersion

$$\delta = \left(\frac{\langle \delta X^2 \rangle}{\langle X \rangle^2} \right)^{1/2} = \frac{(X_b - X_a)}{(X_b + X_a)}$$

Now, let the variable X be *extensive*, that is, proportional to the size of the system N. We express this by setting

$$X = N \cdot x \tag{4.7}$$

where the *intensive variable* x remains finite. This leads to the following estimate of δ:

$$\delta = \frac{(x_b - x_a)}{(x_b + x_a)} = \text{Finite quantity, of order one} \tag{4.8}$$

In other words, the variance is of the same order as the mean, and as a result the effect of the fluctuations reaches a macroscopic level. Clearly we must be violating the premises of the central limit theorem; therefore, the variable distributed according to Figure 69(c) can in no way be the sum of statistically independent random variables. We can already suspect that this should reflect the appearance of *coherent behavior* within the system. An inspection of Figures 69(a) and 69(c) suggests that this coherence must be attributed to a transition from a state of a unique most probable value $\bar{X}$ to a state characterized by two such values, X_a and X_b. This is nothing but the *stochastic analogue of bifurcation*, the phenomenon that was at the center of our preoccupations in Chapter 3. Clearly, in the framework of a deterministic description the problem of how this coherence emerges and is sustained could not even be formulated. This is one of the basic reasons that the study of fluctuations forms an integral part of the dynamics of complex phenomena. Another reason is that in a multihump regime beyond bifurcation, it is important to estimate the relative stability of the most probable states, and the time the system spends in their vicinity before performing a transition to some other simultaneously stable branch. Again, this question can only be formulated in the framework of an enlarged description incorporating the effect of fluctuations.

Figure 69(b) depicts the expected form of the probability distribution in the limit between unimodal and multimodal regimes. We have a flat distribution, which in effect prepares the way for bifurcation by exploring regions of phase space that are quite far from the most probable value. This *probabilistic analogue of marginal stability* will be analyzed further in Section 4.4 and will be shown to lead to a breakdown of the central limit theorem.

So far we have centered our discussion on the role of the fluctuations generated spontaneously by the molecular dynamics of the system. However, as we saw in Sections 3.7 to 3.10, in large classes of dynamical systems random behavior can also arise through the bifurcation of unstable motions leading, for instance, to chaotic attractors. Probabilistic considerations therefore again become an indispensable supplement of the deterministic approach, for they allow us to determine, for instance, the frequency at which different parts of a chaotic attractor will be visited, or the time beyond which the memory of a particular initial state will fade.

The main difference between fluctuations and the randomness associated with chaos is in the *strength* and the *time scale* of the deviations from the average. When a physico-chemical system has a unique asymptotically stable attractor, the fluctuations around the mean value are extremely small, roughly $1/N^{1/2}$ times less than the deterministic values. Even in a multimodal regime in which the dispersion around the mean is large, the dispersion around each of the local peaks remains very small. In addition, the values of the probability in the vicinity of the unstable state are as small as e^{-N}, the inverse exponential of the size parameter N. This means that for most of the time the system stays in the immediate vicinity of a probability peak and that the *transition time* between peaks is very long, of the order of e^N. None of these properties holds for a typical chaotic attractor. In that case the system visits distant states during a lapse of time comparable to the deterministic time scales, indicating that the strength of the deviations from the average is comparable to the average itself. This is reminiscent of the behavior associated with a turbulent motion and underlines once again the role of chaos as a natural model of turbulence in physico-chemical systems.

A similar property characterizes the dynamics of systems subjected to a random environment, whose variability is perceived by the system as an *external noise*. The work by Horsthemke and Lefever listed in the suggested readings is a detailed survey of this important topic.

4.2 MARKOVIAN PROCESSES. MASTER EQUATION

In the preceding section we argued as if the probability distribution describing the system of interest were known, at least as far as its general features were concerned. In some cases this is indeed what happens. For instance, the analysis of Section 2.8 showed that in an ideal system at thermodynamic equilibrium the probability of finding a given number of particles in a small volume is poissonian. Similarly, if in a certain system the validity of the conditions of the central limit theorem given in Section 4.1 is guaranteed, we know that the corresponding variables will be described by the normal distribution. In general, however, all that we are given is the processes that are going on in a dynamical system: heat conduction and convection in the Bénard problem, reactions and diffusion in an isothermal chemical system, and so forth. On the basis of this information, how can we determine the form of the underlying probability distribution?

To address this question it is convenient to set up a balance equation that counts the processes leading the system to a certain state Q and the processes removing it from this state. Obviously,

$$\frac{d}{dt} \text{Prob}(Q, t) = (\text{contribution of transitions to state } Q \text{ per unit time})$$

$$- (\text{contribution of transitions from state } Q \text{ per unit time})$$

$$\equiv R_+(Q) - R_-(Q) \tag{4.9}$$

and the problem reduces to determining the transition rates R_+ and R_-. Whatever the explicit form of these quantities may turn out to be, it will have to be compatible with certain constraints. For instance, it is known that at thermodynamic equilibrium physico-chemical systems satisfy the conditions of *detailed balance*, to which we alluded repeatedly in Chapters 1 and 2. Thus, if we decompose R_+

and R_- into the elementary processes taking place in the system,

$$R_\pm = \sum_k r_{k,\pm}$$

the following condition must be satisfied:

$$(r_{k,+})_{equil} = (r_{k,-})_{equil} \tag{4.10}$$

These relations must in turn be compatible with the form of the probability distribution in the state of equilibrium, which is known from statistical mechanics. A limiting case of such a distribution is the poissonian. More generally, as shown first by Einstein, at equilibrium the probability of a fluctuation is determined entirely by thermodynamic quantities. In an isolated system, the inversion of the famous Boltzmann formula

$$S = k_B \ln \text{ (number of molecular arrangements compatible with a given}$$
$$\text{value of energy)}$$

yields:

$$P_{eq} \sim \exp \frac{1}{k_B} \Delta S \tag{4.11}$$

where ΔS is the change of entropy due to the fluctuation, $\Delta S = S(Q) - S(Q_{equil})$.

Another constraint Eq. (4.9) should satisfy is to reduce, in a certain limiting sense, to the evolution equations we deal with in the deterministic description, such as the equations of fluid mechanics and chemical kinetics. We expect that a macroscopic observation of a physico-chemical system will yield values representative of its most probable state. Therefore we must require that the peaks of $P(Q, t)$ be solutions of the deterministic equations. If the distribution is unimodal [Figure 69(a)], this in turn implies that the equation for the mean value is close to the deterministic equation, the correction being essentially proportional to an inverse power of the size of the system.

Despite their interest, the above constraints are not sufficient to fix the form of Eq. (4.9) in a unique fashion. We shall therefore look at some intuitive arguments. We have stressed that fluctuations arise from the random motion of molecules, which is due to the intermolecular interactions and to the large number of particles involved in the dynamics of most physico-chemical systems of interest. They are therefore essentially localized events whose characteristic space and time scales are extremely short. Under these circumstances we expect that the transition rates appearing in Eq. (4.9) will depend only on the state Q and on the states that can be connected to Q through a single fluctuation. In other words, we suppose that the memory of successive transitions arising from successive fluctuations is lost beyond the first transition. This condition defines an extremely important class of phenomena known as *markovian processes*, whose study is one of the main preoccupations of modern probability theory.

A very simple illustration of a markovian process is shown in Figure 70. The system considered has five states, I to V. The end states, I and V, are "absorbing boundaries" in the sense that the system stays there once it reaches one or the other. But once in state II, III, or IV, the system jumps immediately to the left or to the right with probabilities q and $1 - q$, respectively. In the limit where the number of states is very large, this "random walk" becomes a realistic representation of the motion of a particle in a host fluid and leads to the familiar Fick law of diffusion.

In systems presenting complex behavior we deal with more intricate situations because of the nonlinearities of the kinetics. Nevertheless, an explicit form of Eq. (4.9), known as the *master equation*, can be written. It expresses the rate $R_+(Q)$ as the product of the transition probability per unit time of going from state Q' to Q, times the probability of being in the state Q' at time t in the first place, summed over all states Q' that can lead to Q in a single step by virtue of the elementary dynamical processes going on in the system. Similarly, $R_-(Q)$ is the product of the probability of being in state Q at time t in the first place, times the sum of the transition probabilities per unit time from Q to all states Q' accessible from Q. We write this balance in the form:

$$\frac{dP(Q, t)}{dt} = \sum_{Q' \neq Q} [W(Q|Q')P(Q', t) - W(Q'|Q)P(Q, t)] \qquad (4.12a)$$

where the transition probability per unit time $W(Q|Q')$ is nonnegative for any $Q' \neq Q$, is assumed to be time-independent (stationary Markov process), and must satisfy the condition

$$\sum_{Q} W(Q|Q') = 0 \qquad (4.12b)$$

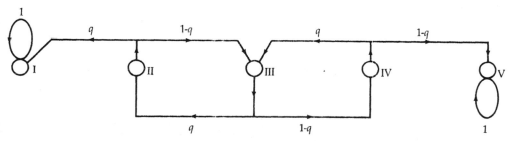

Figure 70 Illustration of a markovian process. The outgoing arrows from states I to V indicate the direction in which the allowed transitions take place.

which expresses the intuitively clear fact that, starting from state Q', all the system can do is either stay in Q' or perform a transition to one of the accessible states $Q \neq Q'$.

Equation (4.12a) is linear with respect to the unknown function $P(Q, t)$, but this does not mean that its solution is straightforward. In fact, the complexity of the dynamical laws of evolution (for instance, the feedbacks present in a chemical or biological system) is reflected by the presence of nonlinearities in the transition rates W. Near equilibrium these feedbacks are ineffective because of detailed balance, but far from equilibrium they are fully manifested. As a result the complete analysis of Eq. (4.12a) constitutes an open problem. Still, by carrying out perturbative calculations, by looking at simple, exactly soluble limits, and by carrying out computer simulations we arrrive at a general picture of the laws governing the fluctuations in nonlinear, nonequilibrium systems. The main results are presented in the subsequent sections of this chapter. Here we examine a few concrete illustrations of the structure of the master equation.

A class of phenomena amenable to Eq. (4.12a) is composed of *birth and death processes*. These are phenomena in which the state Q is characterized entirely by a set of integers $\{X_\alpha\}$, which can either remain unchanged or vary by only 1 or -1. Obvious examples of such a phenomenon are chemical and biochemical reactions, in which case $\{X_\alpha\}$ represent the number of molecules of species α present. Considering for simplicity a particular species X, we can write

$$W(X|X') = \sum_\rho W_\rho(X|X - r_\rho)\delta^{Kr}_{X', X-r_\rho} \qquad r_\rho = 0, \pm 1 \qquad (4.13a)$$

where ρ labels a particular reaction step and r_ρ is the number of molecules consumed ($r_\rho < 0$) or created ($r_\rho > 0$) in the process. We define the birth and death rates, respectively λ and μ, as follows:

$$\lambda_X = \sum_\rho W_\rho(X|X - 1) \equiv N\lambda(X/N) \qquad X \geqslant 0$$

$$\mu_X = \sum_\rho W_\rho(X|X + 1) \equiv N\mu(X/N) \qquad X \geqslant 1$$

$$(4.13b)$$

The last part of the above equalities emphasizes one important feature of the stochastic dynamics: Because of the short-range character of the interactions, the number of transitions per unit time going on in a system of size N must be proportional to the size. That is, each small element of the system can only sense its neighbors; therefore, by summing over all elements we obtain a factor N times a function of the intensive variable X/N. Naturally, we assume that we are dealing with interactions of physico-chemical origin. In the dynamics of social systems the situation might be different. Because of telecommunications, mass media, and similar mechanisms, each element of such a system can simultaneously sense all other elements present. This would lead to transition rates proportional to some power N^a, where the exponent a may be larger than unity.

Let us illustrate the structure of the transition probabilities per unit time on the autocatalytic model of Eq. (2.15). We must evaluate the four quantities W_1, W_{-1}, W_2, W_{-2} describing, successively, the forward and backward rates of the two chemical reactions involved in the model. Consider first W_1. The reaction $A + 2X \xrightarrow{k_1} 3X$ arises from the encounters of pairs of X particles with particles of species A. Therefore, invoking combinatorial arguments for the number of collisions, we have

$$W_1(X + 1 | X) = k_1 \frac{AX(X - 1)}{N^2} \qquad (\text{step } A + 2X \xrightarrow{k_1} 3X)$$

where the factor $1/N^2$ ensures that the whole expression is extensive. Similarly,

$$W_{-1}(X - 1 | X) = k_2 \frac{X(X - 1)(X - 2)}{N^2} \qquad (\text{step } 3X \xrightarrow{k_2} A + 2X)$$

$$W_2(X - 1 | X) = k_3 X \qquad (\text{step } X \xrightarrow{k_3} B)$$

$$W_{-2}(X + 1 | X) = k_4 B \qquad (\text{step } B \dashrightarrow{}^{k_4} X)$$

The birth and death rates are given by

$$\lambda_X = W_1 + W_{-2}$$

$$\mu_X = W_{-1} + W_2$$

Summarizing, from Eqs. (4.13) and (4.12) we can write the following form of the master equation for a birth and death process:

$$\frac{dP(X, t)}{dt} = \lambda_{X-1} P(X - 1, t) + \mu_{X+1} P(X + 1, t) - (\lambda_X + \mu_X) P(X, t) \quad (4.14)$$

It is instructive to check, in this fairly explicit form, the connection between stochastic and macroscopic descriptions. Multiplying both sides by X and summing over X, we obtain from the stochastic description

$$\frac{d\langle X \rangle}{dt} = \langle \lambda_X \rangle - \langle \mu_X \rangle$$

whereas the macroscopic description of the same system would yield

$$\frac{d\bar{X}}{dt} = \text{Dominant part of } (\lambda_{\bar{X}} - \mu_{\bar{X}}) \text{ in the limit of large } \bar{X}$$

The two descriptions are therefore different if the average of λ or μ is different from the value of these functions computed at $X = \langle X \rangle$. This is precisely what happens in a nonlinear system. To see the nature of the correction to the macroscopic description, we consider the simple example of a pure death process,

$$2X \xrightarrow{\ k\ } X + E$$

The death rate μ_X counts the frequency of encounters of pairs of X particles. It is therefore given by $\mu_X = k(1/N)X(X - 1)$. We thus have:

$$\frac{d\langle X \rangle}{dt} = -k\frac{1}{N}\langle X(X - 1)\rangle$$

$$= -k\frac{1}{N}(\langle X \rangle^2 + \langle \delta X^2 \rangle - \langle X \rangle)$$

Switching to the intensive variable $x = X/N$, we can write this relation as

$$\frac{d\langle x \rangle}{dt} = \underbrace{-k\langle x \rangle^2}_{\substack{\text{Macroscopic} \\ \text{law}}} \underbrace{-\frac{k}{N}\frac{\langle \delta X^2 \rangle - \langle X \rangle}{N}}_{\substack{\text{Correction due to} \\ \text{the fluctuations}}} \qquad (4.15)$$

We see that the deviation from the macroscopic law is proportional to the deviation of the underlying probability distribution from the poissonian, since in the poissonian distribution $\langle \delta X^2 \rangle = \langle X \rangle$ [see Eq. (2.28)]. This conclusion is in fact quite general and is essentially independent of the details of the kinetics. Coming back to our example, in a system described by a unimodal distribution $\langle \delta X^2 \rangle$ is of the order of N, by the central limit theorem. The correction to the macroscopic law is then weighted by the inverse of the size parameter N and is small if N is large. However, the latter is not necessarily the case. In contrast to the simplifying assumptions leading to the form of Eq. (4.13) for the transition probabilities, the majority of fluctuations are in reality localized events; therefore, N will typically be the size of a region whose characteristic dimension is of the order of the range of the fluctuations. This argument shows that the effect of fluctuations is not universal, but depends crucially on their scale.

The need for a *local approach* to fluctuations goes even further. To ensure the validity of the markovian description, it is essential to choose the state variables in such a way that each individual transition between states can be attributed directly to a single "elementary" fluctuation. But in a global description this will not necessarily be so, and the underlying stochastic process might well become nonmarkovian. In actual fact, in Eq. (4.12a) it is understood that the individual processes, such as chemical reactions, take place in each small volume element ΔV of our system (as in Figure 71), and their effects are subsequently summed over the whole volume V. The various volume elements will of course be statistically coupled through the transfer of particles, momentum, or energy across their boundaries. Thus in our example of the birth and death process the global Eq. (4.14) will have to be supplemented by additional terms describing this process of transfer.

We have argued so far in terms of discrete variables and discrete state spaces. In a physical system this is indeed the fundamental level of description, since fluctuations arise because of the discrete character of matter. In many instances, however, the underlying discrete stochastic process can reduce, in some well-defined asymptotic sense, to a stochastic process with continuous realizations. In this limit the description becomes very similar to the macroscopic one, the difference being that the macroscopic rate is now supplemented with a *stochastic*

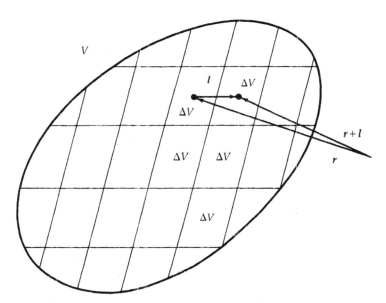

Figure 71 Partition of a macroscopic system into subvolumes of size ΔV centered on space points r coupled to their neighbors $r + l$ through mass, momentum, or energy transfer. Because of the local character of the fluctuations, it is expected that if ΔV is suitably chosen the process of following the evolution of the state variables within the set of all ΔV's will be markovian.

force expressing the effect of the fluctuations. This description, which goes back to Paul Langevin's classic study of brownian motion, proves in many instances to be a fertile approach to the stochastic analysis of large classes of dynamical systems.

4.3 *MARKOVIAN PROCESSES AND IRREVERSIBILITY. INFORMATION ENTROPY AND PHYSICAL ENTROPY*

The reader may have been struck by the apparent similarity between Sections 4.1–4.2 on the one side and Section 3.2 on the other. In both cases a probabilistic language is adopted, and the evolution of the underlying probability distribution is reduced to the solution of a linear equation of first order with respect to time: the master equation, Eq. (4.12a), or the Liouville equation, Eq. (3.7). Despite these analogies between the two descriptions, markovian processes have an altogether different status. This might have been suspected from the relation established in the preceding section between the master equation and the macroscopic rate laws characteristic of dissipative systems, a property that is to be opposed to the time-reversal invariance of the Liouville equation. As we will see, the connection between markovian processes and dissipative systems is in fact a very general feature, one deeply rooted in the very structure of the master equation. Specifically, as time grows to $+\infty$, the solutions of the master equation exhibit a universal tendency toward a unique final stationary state, conditioned by a *Lyapounov function* very similar to the thermodynamic functions considered in Section 2.6.

We begin by introducing the *entropy* of a markovian process. The need for such a quantity arose quite early in probability theory. It was essentially motivated by the development of exactly soluble probabilistic models aiming to reconcile the irreversible trend to equilibrium predicted by Boltzmann's kinetic theory and the reversibility of the laws of motion of conservative systems. Since the advent of information theory, however, entropy has proved to also be an important concept in the mathematical theory of communication. We shall deal more amply with this information theory aspect of entropy later, in Section 4.7. Let us simply observe here that in both physicochemical and information theory contexts, it is customary to seek a quantity that satisfies a number of requirements imposed by intuition and by analogy with thermodynamics; for example,

1. For a given number of states, N, and for

$$P(Q) \geqslant 0, \sum_Q P(Q) = 1$$

the entropy function S takes its largest value for $P_Q = 1/N$.

2. The entropy of a composed system AB equals the entropy of subsystem A plus the (conditional) entropy $S_A(B)$ of subsystem B under the condition that subsystem A is in a given state A:

$$S(AB) = S(A) + S_A(B)$$

This property is usually referred to as *subadditivity*.

3. Adding an event Q_α that is impossible $[P(Q_\alpha) = 0]$ does not change the entropy of the process.

Under these conditions, and provided that S is also required to be a continuous function of the P's, it can be established that S is uniquely determined up to a multiplicative constant:

$$S[P(Q), t] = -\sum_Q P(Q, t) \ln P(Q, t) \tag{4.16}$$

As it stands, this expression is not an adequate Lyapounov function in the most general case of a markovian process describing a system subjected to nonequilibrium constraints. To arrive at such a function we must introduce an additional assumption, namely, that the markovian process at hand possesses a stationary probability distribution, $P_S(Q)$, satisfying identically the master equation, Eq. (4.12a) with $dP_S/dt = 0$. Not all markovian processes are compatible with this requirement. It is known from probability theory that this property is an outstanding feature of stochastic processes whose states are *ergodic*; in other words, all states are visited with probability one and the mean time of returning to any particular state from which the system started (known as *recurrence time*) is finite.

With the help of $P_S(Q)$ we may now construct the quantity

$$K(P(Q), t) = \sum_Q P(Q, t) \ln \frac{P(Q, t)}{P_S(Q)} \tag{4.17}$$

which obviously remains bounded from below and vanishes at $P = P_S$. Its time variation can be computed by using Eq. (4.12a):

$$\frac{dK}{dt} = \sum_{\substack{Q, Q' \\ (Q \neq Q')}} [W(Q|Q')P(Q', t) - W(Q'|Q)P(Q, t)] \ln \frac{P(Q, t)}{P_S(Q)}$$

By performing permutations of the indices Q and Q' and by using the positivity of $W(Q|Q')$ for $Q \neq Q'$ as well as the well-known property of the logarithm, $\ln x \leqslant x - 1$ for any real x, one finally obtains

$$\frac{dK}{dt} \leqslant 0 \tag{4.18}$$

This result, depicted in Figure 72, bears a striking similarity to the results of Sections 2.5 and 2.6 (see especially Figure 25) concerning the monotonic behavior of entropy in an isolated system imposed by the second law of thermodynamics. It should be noted, however, that K is *not* identical to entropy. Moreover, inequality (4.18) applies equallly well in nonisolated systems subjected to permanent nonequilibrium constraints. Still, just as the second law in an isolated system expresses the universal tendency to a unique equilibrium state, inequality (4.18) also expresses that P_S is unique and globally stable, in the sense that any initial condition $P(Q, t)$ is bound to evolve to it as $t \to \infty$. In other words, the problem of instability and multiplicity of solutions, familiar from the study of bifurcations in nonlinear dynamical systems, simply does not arise at the probabilistic level. Figure 69 shows why this is so: Bifurcation is reflected by a qualitative change in the structure of the underlying probability distribution, such as the appearance of multiple humps, rather than by the multiplicity of the probability distribution itself. A notable exception arises, however, in the limit in which the size of the system tends to infinity. The width of each probability peak then becomes zero, and the probability distribution is decomposed in a number of infinitely sharp distributions; an example of this is provided by Eq. (4.6). The various states of the markovian process no longer define an ergodic set, and different initial conditions are attracted to different final probability distributions. This situation can be shown to adequately describe both equilibrium and nonequilibrium transitions in infinitely extended systems.

Another very important consequence of inequality (4.18) is to show that the master equation, Eq. (4.12a), amounts to a *contraction* in the space of the probability distributions since, whatever the details of the dynamics, their "distance" from $P_S(Q)$ is bound to decrease. This property, together with the linearity of the master equation, is at the basis of the deep connection between markovian processes and the important mathematical concept of *dissipative semigroups*. To

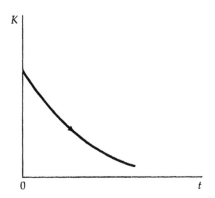

Figure 72 Irreversibility of a markovian process described by the master equation, Eq. (4.12a). The excess entropy function K, Eq. (4.17), decreases monotonically with time.

understand this connection intuitively, let us compare the following two evolution equations:

$$\frac{dy}{dt} = -\lambda y \qquad (t > 0; \lambda > 0) \qquad (4.19a)$$

$$\frac{d^2x}{dt^2} = -\omega_0^2 x \qquad (4.19b)$$

These equations describe respectively, an irreversible decay to a final asymptotically stable state $y = 0$ in a dissipative system (see, e.g., Section 2.2) and a typically time-reversible evolution such as the harmonic oscillator (see Section 3.3). We can also write, instead of Eq. (4.19b), the corresponding Liouville equation in the action-angle variables [see Eq. (3.7) and Section 3.3]:

$$\frac{\partial \rho}{\partial t} = -\omega \frac{\partial \rho}{\partial \varphi} \qquad (4.20)$$

The solutions of the seemingly similar Eqs. (4.19a) and (4.20) are in fact entirely different. For Eq. (4.19a) we simply have

$$y(t) = e^{-\lambda t} y(0) \qquad (4.21a)$$

whereas for Eq. (4.20) a solution compatible with the fact that φ represents an angle must be constructed. This necessitates the introduction of the imaginary unit $i = \sqrt{-1}$:

$$\rho(t) = e^{i(\omega t - \varphi)} \rho(0) \qquad (4.21b)$$

which indeed remains invariant when φ changes by $\pm 2\pi, \pm 4\pi, \ldots$

Both Eq. (4.21a) and (4.21b) display an "evolution operator" acting on the initial condition to produce the instantaneous value. This operator is of the form

$$U_t = \exp(tL) \qquad (4.22a)$$

and satisfies the so-called group property

$$U_t U_{t'} = \exp[(t + t')L] = U_{t+t'} \qquad (4.22b)$$

However, in Eq. (4.21a) $L = -\lambda$ is a negative real number. This implies that U_t is well-behaved only on the positive time axis, in which case we obtain a monotonic decay of the state variable to a unique final steady state. In more abstract terms, we say that the evolution operator U_t defines a *dissipative semigroup*. On the other hand, in Eq. (4.21b) the absolute value of ρ (its "measure" in more abstract terms) remains unchanged: $|\rho(t)| = |\rho(0)|$, and we say that the corresponding operator U_t defines a *unitary group*. Notice that in this latter case the time-reversal invariance of the underlying dynamics is reflected by the property $U_t = U^*_{-t'}$ in other words, time reversal at the level of U_t must be accompanied by complex conjugation.

The above two simple models are conceptually similar to the master equation and the Liouville equation for a conservative system. The only difference lies in the fact that the corresponding L's in the exponential term of Eq. (4.22a) are no longer numbers but operators. We can thus assert that the *signature of irreversibility* lies in the emergence of a dissipative semigroup descriptive of an appropriately defined markovian process. This in turn leads to one of the deepest questions in physics, namely, can such a markovian process arise from a time-reversible dynamics? If the answer to this question turns out to be yes, the transition from a unitary group to a dissipative semigroup would amount to a *time symmetry breaking*. Conservative systems would thus be the "ancestors" of dissipative systems that display this special kind of broken symmetry, just as the uniform steady state of a chemical reagent is the ancestor of the dissipative structure describing a pattern or a rhythm beyond a suitable bifurcation point. Chapter 5 is devoted to a discussion of this fundamental problem.

4.4 SPATIAL CORRELATIONS AND CRITICAL BEHAVIOR

Having discussed the basic features of the probabilistic description of dynamical systems, we can now turn to the applications of these concepts to the various questions raised throughout this book in connection with the emergence of complex behavior.

We have repeatedly referred to the transition phenomena far from equilibrium as being associated with states characterized by correlations of macroscopic range. In Section 2.8 we pointed out that such correlations can only arise in systems in which the regime of poissonian fluctuations has been overcome by the maintenance of suitable nonequilibrium constraints. In Section 4.1 we went even further by associating the coherence entailed by bifurcation with the breakdown of the central limit theorem and the associated enhancement of the fluctuations. In either case, it should be fully realized that the macroscopic coherence and correlations we are talking about have little to do with the intermolecular interaction forces, for they can arise even in a dilute solution of molecules incapable of recognizing each other at a distance greater than a few angstroms. In this section we explain how this rather extraordinary phenomenon can take place, by using the properties of the master equation and by appealing continuously to both the analogies with and the differences from the phenomenon of phase transitions at equilibrium.

As emphasized in Section 1.9, an equilibrium phase transition such as freezing or spontaneous magnetization is the result of a competition between the intermolecular interaction forces, which tend to order the system, and the random thermal motion of the molecules, which has the opposite effect. This is why equilibrium transitions are typically induced by increasing the pressure (which favors the effect of molecular interactions), or by decreasing the temperature (which diminishes the effect of thermal noise). As a matter of fact, even far from a phase transition intermolecular forces confer on a relatively dense system a certain degree of order whose range, however, is short because it is entirely conditioned by the range of the intermolecular forces (see Figure 18). A useful visualization of this situation is provided by Figure 73.

Consider a reference system at thermodynamic equilibrium in which the effect of intermolecular interactions can be neglected, for example, a perfect gas or an ideal solution [Figure 73(a)]. In such a system the fluctuations in a given volume element are poissonian, a property which reflects the complete disorder that prevails as a result of the thermal motion of the constituting particles. Suppose now that attractive interactions are switched on. The particles will tend to form clusters [Figure 73(b)] and this will result in larger deviations of their spatial distribution from the average value, $\langle \delta X^2 \rangle > \langle X \rangle$. If, on the other hand, the interactions are repulsive, the particles will be distributed more efficiently within the volume [Figure 73(c)] and this will diminish the deviation from the average. In both cases the spatial "order" associated with the nonpoissonian behavior remains short-range, unless the system is driven near a critical point of phase transition: the spatial correlation then acquires a long tail, reflecting the ability to undergo a change encompassing the entire system.

Consider now an ideal chemically reactive system capable of functioning far from as well as close to equilibrium. For such a system we would expect at first sight that the effects of clustering and critical behavior described above will not

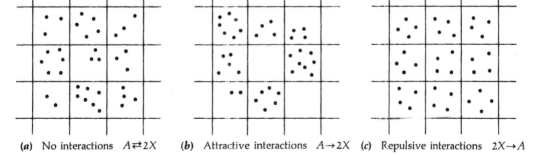

(a) No interactions $A \rightleftarrows 2X$ **(b)** Attractive interactions $A \rightarrow 2X$ **(c)** Repulsive interactions $2X \rightarrow A$

Figure 73 Illustration of the origin of spatial correlations of macroscopic range in a nonequilibrium ideal system. In equilibrium (a) detailed balance precludes accumulation of X in preferred regions of space. X is thus distributed randomly (Poisson statistics), just as one would expect in a system of noninteracting particles. Far-from-equilibrium detailed balance is not valid. Pairs of X can thus be created (b) or destroyed (c) in a correlated fashion. These correlations will be maintained on a distance of the order of the diffusion length. The system, though ideal, will thus behave like a system of particles interacting through attractive (b) or repulsive (c) long-range forces.

take place under any circumstances. In equilibrium this can be checked explicitly: the probability distribution is poissonian, and as a result spatial correlations do not exist. But what happens if the system is driven far from equilibrium by means of appropriately applied constraints? To be specific, consider the reaction $A \rightleftarrows 2X$ and suppose first that A is in large excess, so that the step $2X \rightarrow A$ can be neglected. To avoid accumulation of X to infinite concentrations, we also introduce a reaction removing it from the medium. This leads us to the scheme

$$ A \xrightarrow{k_1} 2X \qquad X \xrightarrow{k_2} B \qquad (4.23) $$

Clearly, the first step amounts to creating a cluster of particles of the species X in the system. In other words, the nonlinear chemical kinetics gives rise to an effect reminiscent of attractive interactions. In equilibrium, the property of detailed balance ensures that any reaction, such as $A \rightarrow 2X$, will be in the average as frequent as its reverse, $2X \rightarrow A$. The clustering described above will therefore be counteracted by the fact that $2X \rightarrow A$ depletes the system of pairs of particles, thus playing a role analogous to repulsive interactions. In this way the familiar poissonian behavior will be recovered. Similarly, if all reaction steps were linear, particles would be created or destroyed at random places in space, and as a result poissonian behavior would still prevail. The presence of both *nonlinear kinetics* and *nonequilibrium constraints* will disturb this balance, and as a result a systematic deviation from the Poisson law will be sustained. How can we relate these deviations to the existence of spatial correlations? In the step $A \rightarrow 2X$ the two X particles are obviously correlated, as they are created together. Owing to diffusion, this correlation will spread in space until at least one of the particles becomes inactivated through the second step of the scheme, Eq. (4.23). The resulting correlation length is thus given by the estimate:

Rate of inactivation of X through the second step of Eq. (4.23)
$\simeq$ Diffusion rate over a distance equal to the correlation length

that is

$$ k_2 x \sim \frac{D}{l_{corr}^2} x $$

or

$$ l_{corr} = \sqrt{\frac{D}{k_2}} \qquad (4.24) $$

in which D is Fick's diffusion coefficient of species X.

Note that if we denote by $F(X)$ the rate law corresponding to scheme (4.23), $F(X) = k_1 A - k_2 X$, then k_2 is simply equal to $|F'(X_s)|$ which, according to Section 3.6 and Appendix 1, gives the rate of decay, γ, of the perturbations around the steady state $X_S = k_1 A / k_2$. This conclusion is in fact quite general. In any system undergoing a transport process whose rate is described by a coefficient D and a local "chemicallike" process whose slowest mode is characterized by the decay rate γ, the correlation length is given by

$$l_{corr} \simeq \sqrt{\frac{D}{\gamma}} \qquad (4.25)$$

As we saw repeatedly in Chapter 3, in the vicinity of a bifurcation point γ tends to zero, and as a result the correlation length diverges. This in turn indicates that coherence tends to encompass the entire volume of the system, which thus becomes able to collectively undergo a transition to a new state. Note however that because of the slowing down of the evolution as it approaches criticality, near the bifurcation point, it becomes impossible to argue in terms of individual modes and their decay rates. A more complete theory taking into account the *nonlinear coupling* of modes is needed. Such an approach has been worked out using both the full master equation and its continuous limit. The main qualitative conclusions are that near the bifurcation point the law of divergence of the correlation length is modified with respect to Eq. (4.25) because of the mode coupling, and that the transition may even be suppressed altogether.

To understand this latter, most surprising point, it is necessary to introduce the concept of *critical dimensionality*. Remember that we are dealing with systems in which a local process (for instance, an autocatalytic chemical reaction) coexists with a transport process and that these processes couple neighboring volume elements among themselves (see Figure 71). Because of fluctuations, the chemicallike process will tend to produce deviations from spatial homogeneity. In the extreme case of no coupling between volume elements, this would lead to a random combination of localized states: even if within each element the parameters are close to their bifurcation values, the system will simply not "see" the bifurcation because it will be dominated by noise. On the other hand, in the opposite case in which the space cells are coupled but the chemicallike process is absent, spatial homogeneity will be secured but order will again be absent since there will be simply no "message" to be relayed and sustained. We conclude therefore that both types of processes are necessary and should be of comparable importance or, in more concrete terms, that they should both have comparable characteristic times. Now, the characteristic time for smearing an inhomogeneity through transport over the length L of a spatial cell is simply given by

$$\tau_D \simeq (D/L^2)^{-1}$$

or, using the relation $L^d = \Delta V$ where d is the number of space dimensions in which the system is operating:

$$\frac{1}{\tau_D} \simeq D \, \Delta V^{-2/d} \qquad (4.26)$$

If spatial coherence is established over a macroscopic range, ΔV will be large, and hence diffusion will be slow. Moreover, this slowing down will be more important as the dimensionality becomes smaller. On the other hand, close to a bifurcation point the chemicallike process will experience a critical slowing down, as a result of which its characteristic time τ_{ch} will also become large. However, in contrast to τ_D the slowing down will be independent of the dimensionality, because of the local character of the process. In fact, for a given cell ΔV it will be given by the inverse of the decay rate, γ, of the slowest mode and will thus depend only on the parameters and the degree of the dominant nonlinearities. By requiring that both τ_D and τ_{ch} be of the same order, we arrive at a relation linking the value of dimensionality to the intrinsic properties of the system: this is precisely the critical dimensionality, d_c. For $d \gtrsim d_c$, diffusion will be efficient, and as a result long-range spatial correlations will be sustained. But if d is well below d_c, diffusion will not be able to correlate the various spatial cells sufficiently well, and bifurcation will be "missed."

The specific value of d_c depends rather weakly on the detailed structure of the system, since what seems to matter are some quite general features such as the degree of the dominant nonlinearity. We can define "universality classes" in this way—classes constituted of quite different systems which nevertheless show similar qualitative behavior. For instance, in both the autocatalytic model of Eq. (2.15) and in the brusselator, Eq. (3.18), the critical dimensionality for certain bifurcations turns out to be $d_c = 4$. Note however that far from the bifurcation the dimensionality ceases to play an important role. For instance, the estimate of Eq. (4.25) for the correlation length far from bifurcation remains valid whatever the value of d.

To go beyond these qualitative results we have to construct the explicit form of the solutions of the master equation, Eq. (4.12a). We will not discuss the technical aspects of this problem here; rather, we outline some of the highlights:

1.   As soon as a system is not strictly at equilibrium, correlations of a macroscopic range suddenly arise and can be sustained indefinitely. In systems in which a chemicallike mode is coupled to a transportlike mode, the range is intrinsic and is given essentially by Eq. (4.25). The precise form of the correlation depends on both the geometry and the intrinsic parameters. In a three-dimensional system of large spatial extension, we thus find a correlation function between locations r and r' [closely related to the covariance, see

Eq. (4.2)] of the form

$$g(r - r') = \frac{J_s}{2\pi D} \frac{1}{|r - r'|} \exp\left[-\left(\frac{|F'(X_s)|}{D} \right)^{1/2} |r - r'| \right] \qquad (4.27)$$

whose amplitude J_s is proportional to the strength of the nonequilibrium constraint. In a sense, therefore, the amplitude J_s can be considered as an order parameter that characterizes the transition from an uncorrelated state of matter (equilibrium) to a correlated one (nonequilibrium). Notice that the correlation length itself remains finite in equilibrium. On the other hand, in systems in which there is no coupling of the kind mentioned above between different modes, the correlation length may be extrinsic. For instance, in a heat-conducting medium at rest subject to a temperature difference, we find that the correlation length is essentially given by the size of the system. The amplitude, however, remains related to the strength of the nonequilibrium constraint (here, the energy flux through the system) and again plays the role of order parameter. The situation is described qualitatively in Figure 74.

2. When a bifurcation point is approached, the correlation length diverges, provided that the dimensionality is not too low. Correlations therefore extend throughout the system. This gives rise to a qualitative change of the properties of the underlying probability distribution that is somewhat similar to the change between Figure 69(a) and 69(b), the difference being that we now deal with multivariate distributions depending on the values of the stochastic variables X at each point of space r. Specifically, in an infinitely extended system, such as the autocatalytic model of Eq. (2.15), undergoing a bifurcation to multiple steady states

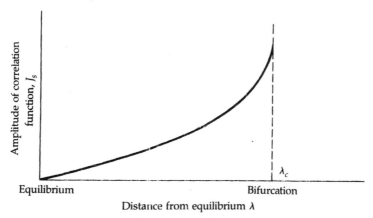

Figure 74 As soon as a system is not at equilibrium, space correlations of macroscopic range emerge whose amplitude is an increasing function of the nonequilibrium constraint.

without spatial symmetry breaking, it turns out that the stationary probability distribution near the cusp singularity [see Figure 42(a)] has the form

$$P_S(\{x_r\}) = \exp\left\{\frac{2\,\Delta V}{Q_S}\sum_r\left[\frac{(\lambda-\lambda_c)}{2}(x_r-x_S)^2 -\tfrac{1}{4}(x_r-x_s)^4\right.\right.$$
$$\left.\left.+\frac{\mathscr{D}}{8d}\sum_l(x_{r+l}-x_r)^2\right]\right\} \tag{4.28}$$

Here $\lambda - \lambda_c$ is the distance from bifurcation; $\mathscr{D}$ is the jump rate between neighboring volume elements ΔV related to Fick's diffusion coefficient D; Q_S is a certain positive combination of the system's parameters; and x_r is the local value of the state variable X_r reduced by the volume element ΔV ($x_r = X_r/\Delta V$). The index l labels the neighbors of cells r to which the exchange of X is limited through the transport processes present in the system.

The nongaussian character of this distribution reflects the coherence associated with bifurcation: the system can no longer be partitioned into a collection of weakly coupled subsystems. Similar results are obtained for bifurcations leading to limit cycles and to spatial structures associated with symmetry breaking.

An interesting manifestation of the breakdown of the central limit theorem is the dependence of the variance of the fluctuations of the particle number in a finite volume, on the size V. The evaluation of this quantity from Eq. (4.28) in the regime of marginal stability, at which the transition to a new state takes place, leads to:

$$\langle\delta X^2\rangle_V \sim V^{3/2}$$

This is to be opposed to the dependence of $\langle\delta X^2\rangle_V \sim V$ that characterizes the gaussian regime prevailing well below bifurcation.

The exponential term in Eq. (4.28), known as the Landau-Ginzburg potential, is very familiar from the theory of phase transitions at equilibrium. Using this analogy we can compute many relevant properties of the system, such as the range of correlations, the variance of the fluctuations, and others, using the *renormalization group* techniques developed by Kenneth Wilson.

In short, we arrive at a simple, appealing picture of how order can emerge in a system. In somewhat anthropomorphic terms, order appears to be a compromise between two antagonists: the nonlinear chemicallike process, which through fluctuations sends continuous but incoherent "innovation signals" to the system; and the transportlike process, which captures, relays, and stabilizes the signals. Disturbing the delicate balance between these two competing actors leads to such qualitative changes as an erratic state in which each element of the system acts on its own; or, on the contrary, a "homeostatic" fossillike state in which fluctuations are crushed and a full uniformity is imposed.

Complexity therefore appears to be limited on opposite sides by different kinds of states of disorder. It is difficult to avoid the feeling that this conclusion, which follows from quite general laws governing the dynamics of physico-chemical systems, should provide valuable clues for modeling systems that are beyond the realm of physics and chemistry. Chapter 6 is devoted to this problem of the transfer of knowledge.

4.5 TIME-DEPENDENT BEHAVIOR OF THE FLUCTUATIONS. THE KINETICS AND TIME SCALES OF SELF-ORGANIZATION

Toward the end of Section 4.1 we identified the two principal reasons motivating the study of fluctuations in nonlinear dynamical systems. The first reason, namely, the mechanisms by which coherence emerges and is sustained, led to the analysis in Section 4.4. The second, namely, the kinetics of the transitions between states, constitutes the subject of this section.

Our starting point is Eq. (4.15), which we rewrite in the more general form

$$\frac{d\langle x \rangle}{dt} = \text{(Macroscopic rate law)} + \text{(deviation from poissonian)} \qquad (4.29)$$

As we saw previously, when the bifurcation point is encountered the correlation function and hence the deviation from the poissonian undergo a radical change because of the divergence of the characteristic length. At this stage, therefore, time-dependent solutions of the master equation, Eq. (4.12a), must be envisaged to describe the growth of the critical mode in the vicinity of and above the instability. As this growth goes on, the deviation of the variance from the poissonian value becomes increasingly marked. Thus, the second contribution to the righthand side of Eq. (4.29) may be nonnegligible in certain volume elements ΔV of small size and, after a sufficient (possibly very long) lapse of time, it can take over and drive the average to a new macroscopic regime.

Figure 75 illustrates this *nucleation theory view* of nonequilibrium transitions, originally due to the authors. The system studied is the autocatalytic model of Eq. (2.15). Initially we are given a probability distribution peaked around one of the (deterministically) stable branches arising beyond bifurcation. There are three alternative views of this situation.

In Figure 75(a), which is to be compared with Figure 42(b), the multiple steady states are plotted against a characteristic parameter μ—a suitable combination of the rate constants and the control chemical variables A and B. The upper and lower branches, x_+ and x_-, respectively, are asymptotically stable, whereas the middle branch, x_0, is unstable. Figure 75(b) shows the plot of a useful quantity,

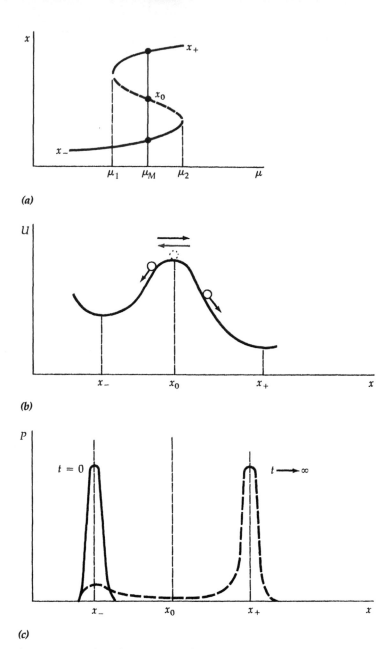

Figure 75 Fluctuation-induced transitions between simultaneously stable steady states. (a) Dependence of solutions of deterministic equations on the parameter μ. It is assumed that at $\mu = \mu_M$ state x_- is less stable than state x_+. (b) Kinetic potential U, Eq. (4.30), as a function of the state variable x for $\mu = \mu_M$. Random fluctuations around each of the stable attractors x_+, x_- are damped unless they lead the system to the other basin of attraction across the potential barrier at the unstable state $x = x_0$. The greater depth of U at $x = x_+$ reflects the assumption that x_+ is more stable than x_-. (c) A probability distribution initially peaked at $x = x_-$ will eventually tend to a bimodal form whose hump at x_+ is much larger than the hump at x_-. The time scale of this evolution is given by Eq. (4.31).

the *kinetic potential*, which was briefly treated in Section 2.6. This quantity is defined through the relation

$$\frac{dx}{dt} = -\frac{\partial U(x)}{\partial x} \qquad (4.30)$$

In systems involving only one variable, U always exists and is equal to minus the integral of the rate law: $U = -\int dx\, F(x)$. In multivariable systems, on the other hand, it is in general impossible to guarantee the existence of a kinetic potential. In any case, whenever it can be constructed the potential proves to be a useful quantity, as its minima and maxima correspond respectively to the asymptotically stable and unstable solutions of the phenomenological equations of evolution. Finally, the solid-line curve in Figure 75(c) represents the initial probability distribution centered on state x_-.

Let us now describe the main results. Suppose that the system of Figure 75 can be maintained spatially uniform within a volume element ΔV containing N particles of the active chemical species responsible for the bistability. From our previous discussion we expect that the characteristic dimension of ΔV will be related to the correlation length. A simplified description of this situation can be based on the solution of the master equation of a homogeneous system of size N, in which diffusion is neglected. We then find a transition to the second stable branch, x_+, of Figure 75(a) whose time scale is determined by N and by the difference of the values of the potential U between the initial state x_- and the unstable state x_0. The final probability distribution is represented by the broken-line curve in Figure 75(c) for the case in which the minimum of U at the initial state is much shallower than the minimum of x_+.

The mechanism responsible for this phenomenon is closely related to the activation energy theory of chemical kinetics, mentioned briefly at the end of Section 1.4. Because of the fluctuations, the different parts of the system experience spontaneous, random deviations from the initial most probable state, x_-. The majority of these fluctuations have a small intensity and range, but finite range fluctuations will also occur with a certain probability. Some of these fluctuations will drive the subsystem beyond the "potential barrier" constituted by the unstable state x_0. At this point the subsystem will jump quickly to the other attracting state, x_+, thereby constituting a "nucleus" of instability. Since by assumption the state x_+ is more stable than x_-, the probability of escaping from it will be small; in addition, new elements of the system will experience similar transitions, and eventually the entire system will become contaminated.

Detailed calculations show that the typical time scale for the transition from x_- to x_+ is of the order of:

$$\tau_{trans} \simeq \exp N\,\Delta U \qquad (4.31)$$

where $\Delta U = U(x_0) - U(x_-)$ is the "potential barrier" that has to be overcome by the fluctuations. If a physico-chemical system of large spatial extension for which

$N \sim 10^{23}$ is artificially kept uniform—say, by an extremely efficient stirring—τ_{trans} will be extremely large, much larger than any conceivable geological or cosmological time. This shows that in real-world situations the transition has to start as a localized event, involving an element of small size. Even so, the characteristic time τ_{trans} will remain considerably larger than the characteristic relaxation times displayed by the deterministic laws of evolution unless the system is very close to the turning points μ_1 and μ_2 [Figure 75(a)], at which the value of the potential barrier becomes very small.

The need expressed above for a local description implies that we must resort to a more detailed theory based on a multivariate master equation that includes the effects of diffusion (see Figure 71). Few analytical results are available; however, numerical simulations establish that the time scale of the transition depends in a highly interesting way on the value of the diffusion coefficient. When this coefficient is small compared to the chemical rate, we find that transitions from x_- to x_+, including the appearance of small "nuclei" of state x_- within the state x_+, occur at a high rate. Subsequently, these nuclei grow slowly and by diffusion entrain the remainder of the system to the new branch of solution x_+. On the other hand, when the diffusion coefficient is large we find that the initial state, x_-, is maintained for a substantial amount of time. This is reminiscent of *metastability*, a property familiar from equilibrium phase transitions. As shown by Roux et al. in experiments on the chlorite-iodide reaction, a similar behavior is observed when the system is subjected to *external* inhomogeneous perturbations caused by, for instance, deviations from completely effective stirring.

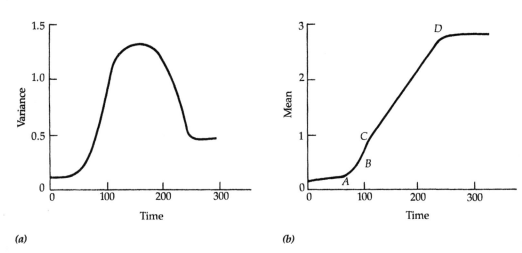

(a)   (b)

Figure 76 Nucleation-mediated transition between multiple steady states in a model system. (a) Fluctuation variance. (b) The time evolution of the mean accelerates sharply as soon as a germ of state x_+ appearing initially through a fluctuation begins to invade a substantial part of the system, initially in state x_-.

Figure 76 describes the kinetics of the fluctuations in the transition between multiple steady states. The variance of the fluctuations is shown in (a), and the mean value in (b). After a relatively long waiting time, a first germ emerges— point A on the curve in Figure 76(b)—and a second germ follows at point B. At point C the two germs reach the boundaries of the system almost simultaneously, whereupon two converging wavefronts are formed and eventually disappear when state D, corresponding to the branch x_+, is reached. Note the appreciable enhancement of the fluctuations around the inflection point of the mean value versus time curve, as witnessed by the abrupt maximum of the variance.

In addition to the above-described phenomena, rich transient behavior is also expected in systems in which a stage of slow motion is followed by an evolution occurring on a much faster time scale. One possible instance of this behavior is when the system is initially prepared at an unstable or a marginally stable state. A second possibility, to which we restrict ourselves here, arises when an initial *induction regime* characterized by a very small rate of change of the pertinent variable is suddenly interrupted by a violent *explosive behavior*, ignited at some characteristic time t_c. This is illustrated by Figure 77. Eventually the system is stabilized at some final state, which for simplicity is assumed to be the only stable state available in the range of interest of the parameter values. A typical example in which these conditions are realized is combustion; other example include auto-catalytic chemical reactions and switching phenomena in lasers.

Let us illustrate the main features of this kind of evolution with an example of adiabatic explosion, that is, combustion in a closed vessel in which the heat released by the chemical reactions is disposed of in heating the mixture. The simplest nontrivial case is a single irreversible exothermic reaction

$$X \xrightarrow{k(T)} A \qquad (4.32)$$

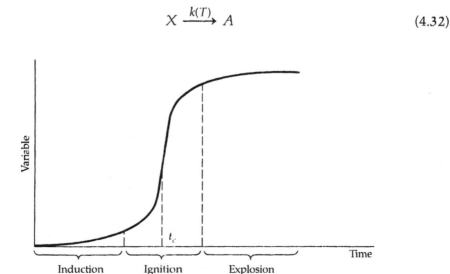

Figure 77 Illustration of explosive behavior.

whose rate constant $k(T)$ depends on temperature T through the Arrhenius law, $k(T) \simeq k_0 e^{-E_0/k_B T}$ [see Eq. (1.5)]. The mass and energy balance equations are thus

$$\frac{dx}{dt} = -k(T)x$$

$$C_v \frac{dT}{dt} = -r_v \frac{dx}{dt} = r_v k(T)x \tag{4.33a}$$

where C_v, r_v are respectively the specific heat and the reaction heat at constant volume. Multiplying the first equation by r_v and adding, we easily check that the following conservation condition is satisfied at all times:

$$C_v T_0 + r_v x_0 = C_v T + r_v x \equiv C_v T_{max} \tag{4.33b}$$

in which T_0, x_0 are the initial values of T, x, and T_{max} is the final temperature attained when the reaction has been completed. Because of this conservation law Eqs. (4.33a) become closed. For instance, the heat balance equation simply reads

$$\frac{dT}{dt} = (T_{max} - T) \cdot k(T) \tag{4.33c}$$

Figure 78 describes the solution of Eqs. (4.33). It is seen that the reaction rate reaches its maximum value abruptly at a time t_c, which will be referred to as the "explosion time," in agreement with the general picture of Figure 77.

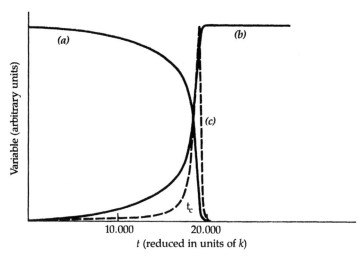

Figure 78 Solution of Eqs. (4.33). Curve a, solution for concentration, x; curve b, solution for temperature, T; curve c, solution for rate of reaction. Parameter values: $E_0/R = 10{,}000$; $T_{max} = 2000°$; $r_v/C_v = 1200$.

Let us now incorporate the fluctuations into the description. Clearly we are dealing with a pure death process, since the concentration of X can only decrease. We can therefore write a master equation in the form of Eq. (4.14) in which the birth rate λ is set equal to zero. Denoting by X the number of particles of the reactive species, we have

$$\frac{dP(X, t)}{dt} = \mu_{X+1} P(X + 1, t) - \mu_X P(X, t) \tag{4.34a}$$

The death rate can be readily computed thanks to the conservation equation, Eq. (4.33b):

$$\mu(X) = k(T)X = k_0 X \exp \left[-\frac{E_0}{k_B \left(T_{max} - \dfrac{r_v}{C_v} \dfrac{X}{N} \right)} \right] \tag{4.34b}$$

where N is a parameter proportional to the size of the system.

Figure 79 depicts the main stages of the evolution of the probability distribution predicted by Eqs. (4.34). The initial condition is chosen to be a sharply peaked distribution centered on a state belonging to the flat (induction) part of Figure 77, for which the rate of evolution is predicted to be very small. We then

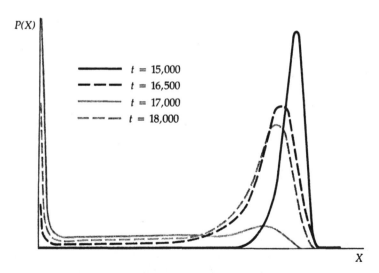

Figure 79 Transient bimodality arising in adiabatic explosion. At $t = 15,000$ time units the probability is unimodal. Beyond $t = 16,500$ time units it develops an increasingly pronounced second peak in the region of low values of the number of particles X. Meanwhile, the amplitude of the initial peak, whose motion follows the slow (induction) time scale is constantly diminishing. Parameter values as in Figure 78.

observe a phenomenon of *transient bimodality* in time: while one of the maxima of the probability distribution remains close to the initial state, a second maximum is formed near the state of complete combustion. Eventually the first peak disappears, and the system is driven to its unique stable state. We therefore have a "bifurcation behavior in time" during the transient evolution of the system.

Let us outline a qualitative explanation of this unexpected phenomenon. Remember that we are dealing with a process involving two widely separated time scales, and that our system is initially prepared in a state in which the deterministic rate is very small. The maximum of the underlying probability distribution, whose motion roughly follows the deterministic one, will therefore move very slowly toward the region of lower values of X. Meanwhile, because of the fluctuations, the probability will develop a width proportional to the length of the induction period and inversely proportional to the square root of N (by the central limit theorem, see Section 4.1). If the length of the induction period is large, the size effect will be counteracted and the width will be appreciable. As a result, a substantial part of the probability mass will reach the ignition point well before the maximum does. At this moment it will be quickly entrained by the fast motion toward the region of low values of X. This leak of probability will go on continuously, but since the system cannot reach negative values of X, a traffic jam will arise as a result of which a new probability peak will emerge in the region of small X. Eventually the primary peak, by then considerably diminished, will reach the ignition point and this will mark the end of transient bimodality.

Obviously, during the lifetime of the transient bimodality the system will display a markedly random behavior: there will be appreciable deviations between the mean and most probable values, and the variance will attain a macroscopic level. This will also show up in the form of fluctuations of the ignition time itself, which will therefore become a random event.

Beyond its specific applications to combustion-related problems, the phenomenon of bifurcations in time provides a new model of differentiation and evolution that have a purely internal origin in nonequilibrium systems. Indeed, the system need not be continuously disturbed for this purpose by the external world—as, for instance, in the darwinian picture of evolution through mutations and natural selection. Rather, the deviations of the dynamics from equilibrium that are created temporarily are sufficient to induce, during some time interval, an internal differentiation: During a slow period of the evolution, fluctuations endow the system with a variety of states visited with some probability, but the initial state remains dominant; the occurrence of a fast stage then entrains some of these states at such a high rate that they soon lose track of the common ancestor, thereby producing a new clone of their own. Similar concepts should also be useful in various problems related to population dynamics and epidemiology or, to come back to physico-chemical systems, in problems involving clustering, collapse, and the proliferation of defects in the process of disorganization or fracture of a material.

4.6 SENSITIVITY AND SELECTION

The analysis in the preceding section provides us with some tools to solve the problem of selection between states, raised repeatedly in Chapter 3. Suppose that a system possesses a number of simultaneously stable solutions. By incorporating the effect of fluctuations, through the master equation, we can compute the heights of the peaks of the probability distribution as well as the average time needed to leave the region of attraction of each of the stable states [see Eq. (4.31)]. The dominant state—the state that will be selected—will then be the state for which the residence time is the longest and around which the probability peak is the highest. Satisfactory as it may sound, this argument leaves one key question unanswered: What happens when the probability peaks and the transition times of at least two of the states are equal?

This possibility should not be dismissed as a pathological case. Some of the most striking transition phenomena we have considered so far correspond to the breaking of symmetries of various kinds and inevitably give rise to multiplets of states that are indistinguishable as far as stability and fluctuations are concerned. One example is the space symmetry-breaking transition at the macroscopic level discussed in Section 3.11. A second example is chiral symmetry breaking at the molecular level, various aspects of which were brought out in Section 3.13. In each case a new selection mechanism is clearly needed. In this section we describe one typical possibility, namely, selection through symmetry-breaking interactions with external fields.

The basis of this mechanism was laid down in the analysis in Section 3.4, which was devoted to the parameter dependence of bifurcation and the related concept of structural stability. As we pointed out in connection with Eq. (3.14), a symmetric bifurcation giving rise to two indistinguishable states [Figure 41(a) and Eq. (3.14) with $\mu = 0$] undergoes a qualitative change when an additional parameter is varied in a suitable way [Figure 42(c) and Eq. (3.14) with $\mu \neq 0$]. Figure 80 summarizes the situation. The corresponding equation for the pertinent variable, x, is

$$\frac{dx}{dt} = -x^3 + \lambda x + \mu \tag{4.35a}$$

and its steady-state solution is given by

$$-x_s^3 + \lambda x_s + \mu = 0 \tag{4.35b}$$

where the parameter μ plays the role of the external asymmetry. Even if its magnitude is small, its effect may be amplified considerably due to the nonlinear nature of the system. The stable branch x_-, which for $\mu = 0$ was indistinguishable from x_+ as far as stability is concerned, still exists but is separated from the

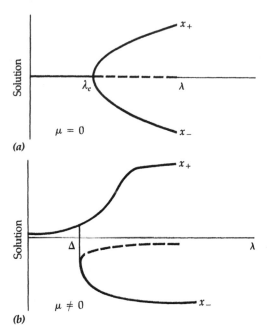

Figure 80 Selection through perturbation of a bifurcation by an external symmetry-breaking field μ. The minimum separation between the perturbed branches, Δ, must be higher than the noise level arising from fluctuations.

x_+ branch when $\mu \neq 0$. Consequently, as the parameter λ is increased from values left of the limit point beyond which x_- exists, the system will tend to stay on the upper branch of Figure 80(b) unless there is a perturbation of sufficient magnitude that knocks the system to the lower branch. Now, since in each system there exist thermal fluctuations, this could happen if the minimum separation Δ indicated in Figure 80(b) is small. Conversely, if the external asymmetry μ is to be a good selector, it should create a separation Δ large enough to overcome the thermal fluctuations. Before we develop this point further, however, we comment briefly on the possible origin of the symmetry-breaking parameter μ.

As emphasized in Section 3.6, equations like Eq. (4.35a) are not limited to a particular model, rather they constitute the typical, *normal form* describing whole classes of dynamical systems functioning near a bifurcation point. Therefore, the question to be asked in connection with parameter μ is how to identify some significant external symmetry-breaking fields which—when coupled to a chemical, hydrodynamic, or other system—give rise to a normal form described by Eq. (4.35a). Let us mention here just two cases, on the grounds of their universality: the gravitational field, which always affects any massive object in the same direction, and the weak-neutral current interactions, which despite their subnuclear origin slightly modify the chemical rate constants of two reactive stereoisomers, also always in the same direction. In both cases we can show that the dynamics can be reduced to a form similar to Eq. (4.35a), provided that in the absence of forcing the system functions in the vicinity of a bifurcation point of the kind shown in Figure 80(a).

We now come to the quantitative estimation of the sensitivity to the small symmetry-breaking parameter μ. From Eq. (4.35b) it is seen that for $\mu \neq 0$ the unperturbed bifurcation point $\lambda_c = 0$ belongs to the region of a unique solution, $x_s = \mu^{1/3}$. On the other hand, the magnitude Δ of the minimum separation can be obtained from the condition that at this point Eq. (4.35b) has three real roots, two of which are coincident. We can derive [see Eq. (3.15)]:

$$\lambda_\Delta = 3(\mu^2/4)^{1/3} \qquad \text{and} \qquad \Delta = \tfrac{3}{2}(4\mu)^{1/3} \qquad (4.36)$$

The presence of the factor $\mu^{1/3}$ in this result is of crucial importance. If the system is functioning close to equilibrium (λ well below zero), the effect of an external field is linear because the cubic terms can be neglected in Eq. (4.35a). Hence the field creates an extremely small bias of the order of $x_s \sim \mu$. But when the system is far from equilibrium and is allowed to fully manifest its nonlinearity, the same interaction can create a much more important bias, $x_s \sim \mu^{1/3}$, leading to an enormous amplification of the signal. For instance, in chiral symmetry breaking, where x_s stands for the difference between right and left stereoisomers, μ is the difference in the activation energy (normalized by the thermal energy) arising from weak interactions. Reliable estimates give $\mu \sim 10^{-15}$. Hence, at equilibrium the bias ($x_s \sim 10^{-15}$) is undetectable at the macroscopic level, but far from equilibrium ($x_s \sim 10^{-5}$) it may begin to compete with macroscopic effects.

As mentioned above, the sharpness of the selection cannot be assessed unless Δ is compared with the internal fluctuations of x in the vicinity of the bifurcation point. The properties of these fluctuations were discussed in Sections 4.1 and 4.4. For our purposes, only some general features are needed.

We first need to specify the deviation of the value λ_Δ at the minimum separation point of the solution from the critical (bifurcation) value $\lambda = 0$. Referring to Eq. (4.28), if λ_Δ is large compared to N^{-1}, where N is a dimensionless measure of the size of the system, the fluctuations around branch x_+ will essentially be gaussian and their variance will be

$$\sigma^2_{Gauss} \sim N^{-1} \qquad (4.37a)$$

But if λ_Δ is very small, then we should compute the variance using the quartic function displayed in Eq. (4.28), which predicts a value of the order of

$$\sigma^2 \sim N^{-1/2} \qquad (4.37b)$$

Sharpness of selection implies $\Delta \gg \sigma$, hence from Eqs. (4.36) and (4.37) we arrive at the condition

$$\tfrac{3}{2}(4\mu)^{1/3} \gg N^{-1/2} \qquad (4.38)$$

This inequality automatically guarantees that $\lambda_\Delta > N^{-1}$. In the problem of chiral symmetry breaking the lefthand side is of the order of 10^{-5}. To estimate the righthand side we take a size of the order of the correlation length [Eq. (4.25)] over which the system remains spatially coherent and consider a dilute solution in which the active stereoisomers are in the range of a millimole. If we can achieve a correlation length of about $l_{corr} \sim 1$ cm near bifurcation, then $N = 10^{18}$ or $N^{-1/2} = 10^{-9}$, and inequality (4.38) is satisfied. Note that near equilibrium the balance would be reversed, since the lefthand side would be replaced by a term of order $\mu \sim 10^{-15}$.

Having seen that far from equilibrium selection is possible in principle, the following prescriptions appear to define a good selection strategy:

1. At $t = 0$, the system is started near $\lambda = \lambda_c = 0$, the bifurcation point in the absence of the field ($\mu = 0$). The distribution is taken to be symmetrical and centered on $x = 0$. As pointed out earlier, in this range there is only one stable state, $x_s = \mu^{1/3}$. The probability peak will therefore drift gradually toward this state. The time scale can be estimated by solving the time-dependent deterministic Eq. (4.35a) with $\lambda = \lambda_c = 0$. By straightforward integration we obtain:

$$\Delta t_{0 \to x} \simeq \frac{1}{3\mu^{2/3}} \left[\ln \frac{\mu^{1/3}}{|x - x_s|} + \text{terms of order one} \right] \qquad (4.39)$$

For x not too close to x_s the order of magnitude of Δt is essentially determined by the factor $\mu^{-2/3}$. Taking again the example $\mu \sim 10^{-15}$, we arrive at $\Delta t \sim 10^{10}$ sec $\sim 10^3$ years! Therefore, the deterministic time scale for selection is exceedingly long.

2. To avoid the difficulty of the above very long selection time, we let the system evolve only until the probability peak has migrated toward a value x_{max} of x around which the noise level (measured by the variance σ) is smaller than x_{max} itself. In view of our previous estimate, for $\mu \sim 10^{-15}$ it would be sufficient to reach a value of x_{max} slightly above 10^{-9}. At this point the system is quenched to a new value of the parameter λ, much larger than λ_Δ. Being now far from critical conditions, the system will evolve exponentially rapidly toward the branch x_+, well before the fluctuations lead to a substantially populated state x_-. Selection is thus expected to be nearly perfect. Numerical simulations show that by this procedure the time scale is shortened to a few days, instead of the 10^3 years predicted by the deterministic equation.

Once again we experience a feeling of astonishment before the amazing potentialities of matter when both nonlinear dynamics and nonequilibrium constraints are present.

4.7 SYMBOLIC DYNAMICS AND INFORMATION

We have described stochastic dynamics in terms of probability distributions and their various moments. A complementary, and for many purposes especially illuminating approach, is the study of individual outcomes of the stochastic process of interest. In this section we summarize some of the basic elements of this approach, which will be utilized further in the following section and in Chapter 5.

Let $\{Q_i\}$ ($i = 1, 2, \ldots$) be the accessible states of a system.* In agreement with the comments made in Section 4.2, it is understood that the set $\{Q_i\}$ is chosen in such a way that the time evolution defines a markovian process. Moreover, instead of considering time as a continuous parameter, we assume that the jumps between the various states occur at instants $t_m = m \, \Delta t$, where m is an integer and Δt has a fixed value depending on the particular process under consideration. This should not constitute a very restrictive assumption, since in nature successive transitions should indeed be distant by some interval of time. Naturally, Δt must be much less than the characteristic time scales of the evolution of the system described by the phenomenological balance equations.

Under the above conditions each possible outcome of a series of ρ consecutive steps of the process can be written as a sequence

$$Q_{i_1} Q_{i_2} \cdots Q_{i_\rho} \qquad (4.40)$$

The set of all sequences of the form of Eq. (4.40) defines a *Markov chain*. Viewing a system as a Markov chain amounts therefore to describing its evolution in terms of the "letters" $\{i_l\}$ of an appropriate "alphabet." At first sight it would seem that for most systems of interest the number of letters in the alphabet should be infinite. This need not be so, however. As an example, consider a system that admits two simultaneously stable steady states. Because of the transitions between states induced by the fluctuations (see Figure 75), the system will spend substantial amounts of time near each of these states—the average time being given by Eq. (4.31)—interrupted by a fast transition to the other state. Let us introduce a more abstract description in which the instantaneous state is labeled "0" or "1," according to whether the system is in the domain of attraction of the first or the second stable steady state. An observer following the evolution by noticing the attraction region in which the system is found at regular time intervals will therefore detect a sequence like

$$111001100010110 \cdots \qquad (4.41)$$

---

* *We first consider a system with a discrete state space. The case of continuous variables is briefly discussed at the end of the present section.*

which provides a description in terms of an alphabet involving only two letters, 0 and 1. In principle, sequences like (4.41) will not define a Markov chain. However, under certain conditions the partition of the initial state space (typically involving a denumerable infinity or even a continuum of states) into the above two states conserves the markovian character. One speaks then of a *markovian coarse graining*. The description afforded by sequences of the type of (4.41) will be referred to as *symbolic dynamics*: We no longer deal with the succession of states in the traditional sense of the term but, rather, with a succession of symbols—the letters of an alphabet of finite length.

Let us try to characterize the abundance of different sequences of the type of Eq. (4.40) or (4.41). From the definition of a markovian process it follows that the probability of the event C described by Eq. (4.40) is

$$P(C) = P_{i_1} p_{i_1 i_2} \cdots p_{i_{p-1} i_p} \tag{4.42}$$

where P_{i_1} represents the initial probability distribution of the state Q_{i_1} and $p_{i_k i_{k+1}}$ the transition probability from state Q_{i_k} to state $Q_{i_{k+1}}$ during the interval Δt. Notice that a given state will typically be visited many times during the process.

We can now prove the following remarkable theorem. Let the process described by the sequences under consideration be markovian and satisfy the law of large numbers (see Section 4.1). The number of all sequences of length ρ involving N states is clearly $N_\rho = N^\rho$. We arrange these sequences in order of decreasing probability $P(C)$ and select from this order the first n_ρ sequences whose sum of the probabilities just exceeds a preassigned positive number between 0 and 1. Then, in the limit of large ρ, the number n_ρ of the sequences so selected depends only weakly on the above preassigned number (which may thus be chosen close to unity), and can be estimated by

$$n_\rho \sim e^{\rho I} \tag{4.43a}$$

where

$$I = -\sum_{ij} P_i p_{ij} \ln p_{ij} \tag{4.43b}$$

The quantity I is customarily regarded as a measure of the amount of information obtained when the Markov chain moves one step ahead of the initial state Q_i. The connection between this quantity and the definition of entropy used in Eq. (4.16) is straightforward. It suffices to consider a deterministic initial probability ($P_i = 1$ if $i = i_1$, $P_i = 0$ otherwise), in which case $p_{i_1 j} = P_j$ represents the probability of being at state Q_j.

The great value of the above theorem can be realized from the following considerations. Whereas the number of all sequences of type (C) is $N_\rho = e^{\rho \ln N}$, the number n_ρ of the selected sequences is approximately $e^{\rho I}$. Now, we may recall (Section 4.3) that the entropy of a Markov chain takes its maximum value when all states are equally probable, $P_i = 1/N$. This gives $I_{max} = \ln N$. Thus we always have $I < \ln N$ (except for a trivial case) and as a result, for ρ large,

$$e^{\rho I} \ll e^{\rho \ln N} \qquad (\rho \to \infty)$$

In other words, only a small specific fraction of all the sequences C has a sum of probabilities close to unity. How small this fraction is depends on the entropy of the Markov chain. We thus arrive at the important conclusion that a deviation from equiprobability acts like an extremely efficient selector of privileged sequences out of the huge set of all possible random sequences. Such a deviation can be realized if the system is not isolated but instead is put in contact with an external environment. As we have seen repeatedly, such conditions can lead to *nonequilibrium states* that can be sustained indefinitely because of the exchange of matter or energy between the system and the external world. Nonequilibrium is therefore the natural environment in which selection may take place.

As an example, consider the process of formation of a biopolymer, say a protein $\rho = 100$ amino acid molecules in length. It is well known that in nature there exist $N = 20$ kinds of amino acids. If biopolymerization were to take place in an isolated system in which all sequences are a priori equally probable, we would have $N_\rho = e^{100 \ln 20} \sim e^{300}$ equally probable sequences. Hence, any particular arrangement—necessary to fulfill, for instance, a desired biological function— would occur with the exceedingly low probability of e^{-300}! If, on the other hand, biopolymerization is performed under nonequilibrium conditions corresponding to an entropy of, say, $I = 0.1 I_{max}$, then only $n_\rho \sim e^{30}$ sequences of length 100 would be realized with an appreciable probability. This number is much closer to our scale. It is thus conceivable that evolution acting on such a relatively restricted population in a nonequilibrium environment can produce, given enough time, entities endowed with special properties such as self-replication, efficient energy transduction, and so on.

So far we have assumed that the dynamical systems giving rise to the stochastic process under study could be defined in a discrete state space in which the evolution was a markovian process. As explained in the first sections of this chapter, these conditions should be satisfied by the internal fluctuations generated spontaneously by the molecular dynamics of all macroscopic physico-chemical systems. But what about chaotic dynamics, the second universal source of stochasticity in nature? To begin with, the state space of systems giving rise to such a dynamics is typically continuous. Moreover, there is a priori no reason for having a markovian process since, for one thing, contrary to the fluctuations, the deviations from the mean motion are neither localized nor small-scale events.

It is remarkable that in many cases chaotic dynamics can, nevertheless, be cast in the form of a Markov chain. The reason is that the state space frequently can be partitioned in a finite number of cells such that, as time goes on, the dynamics induces transitions between these cells satisfying the Markov property. Because of the existence of such partitions a new level of abstraction can be reached in which *symbolic dynamics* involving the succession of letters of an alphabet labeling the various cells again becomes the natural description. In the next section we discuss a particular example of such a symbolic description, in the context of the problem of selection of asymmetric structures.

4.8 GENERATION OF ASYMMETRIC, INFORMATION-RICH STRUCTURES

We have seen how the value of the entropy of a Markov chain conditions the number of sequences that can occur with appreciable probability. Moreover, through the use of symbolic dynamics, it becomes possible to regard these selected sequences as a set of "messages" or "texts" containing information. The question we raise here is how such information-rich sequences can be generated spontaneously by a dynamical system.

When we have a piece of paper containing a text written in English, our comprehension of the message contained therein depends on the fact that we read it from left to right and observe the periods, the commas, and the syntactic rules. A text written in Arabic will make sense only if it is read from right to left, and if the syntactic rules governing the reading of Arabic are observed. Similarly, as pointed out in Section 3.13, the genetic code is characterized by start signals and by instructions allowing it to be read only in a fixed direction from the start point.

A second, equally important aspect of the comprehension of a message is the presence of a particular set of symbols unfolding in time or, more generally, along the direction of reading. Although perfectly well defined and reproducible once known, such a sequence is basically unpredictable in the sense that in all nontrivial cases its global structure cannot be inferred from the knowledge of a part of it, no matter how large. In this respect it can be regarded as a stochastic process, and *it is this recognition that allows us to speak of information.* For instance, what we perceive as information arising from the reading of Euclid's *Elements* or Newton's *Principia* would be very different if we were able to infer by a simple algorithm the second half of these treatises from reading the first half!

In short, information involves two fundamental conditions:

1. A sharp symmetry breaking in space, owing to which other possible issues of the reading process are continuously eliminated (this brings us again to the questions raised in Sections 3.13 and 4.6, namely, how such asymmetric forms can arise and dominate their mirror images).

2. An element of unpredictability, associated with the revealing of an object or a message that the reader could not infer to begin with.

We shall now sketch a mechanism in which both these conditions are automatically fulfilled.

Consider a dissipative dynamical system such as a set of chemical reactions giving rise to the synthesis of species whose concentrations we denote by X_i. Suppose that we record the set of values of some representative variables X_i obtained for increasingly large times. We can regard this sequence as a one-dimensional string of digits. Now, because of the irreversibility inherent in the evolution of any dissipative system, the "direct" sequence so generated will be asymmetric in the sense that typically it will be different from a sequence that we would read in the reverse direction: the latter sequence could in fact only be generated by a physically unacceptable dynamical system resulting from initial time reversal. We therefore see that irreversibility can induce an absolute asymmetry in the abstract space of one-dimensional sequences of symbols. The question we raise next is how to transfer this asymmetry of the symbolic dynamics to a one-dimensional system embedded in the *physical space*.

Let us propose a particular algorithm by which this goal can be achieved. Assume that the chemical system under consideration is capable of producing sustained oscillations in time, oscillations that are not necessarily periodic. Assume furthermore that when X_i crosses a certain level L_i with a positive slope (at the points marked ● in Figure 81) a new process is switched on as a result of which substance i is rapidly precipitating or diffusing outside the reaction space, and is subsequently collected on a "tape." The space pattern that results from this will carry a signature of the sequence of the various thresholds encountered by the various concentrations. As pointed out earlier, this sequence will typically be asymmetric because of the time symmetry breaking afforded by irreversibility. We have thus succeeded in mapping irreversibility onto a one-dimensional spatial

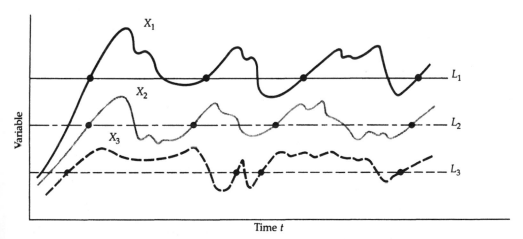

Figure 81 Generation of strings of symbols by a dynamical system. Whenever the state variables X_1, X_2, ... cross levels L_1, L_2, ... with positive slope, a process of precipitation takes place that allows the transfer of this temporal sequence along a spatial coordinate.

pattern. More generally, it can be shown that in a reaction-diffusion system coupled to a *polar* vector (such as the gravitational field or an electrostatic field), space patterns can be generated that carry the signature of the sequence of values reached irreversibly by the concentration variables in the course of time. This spatial mapping of irreversibility cannot be achieved by a coupling with an axial vector such as a magnetic field, since such a coupling cannot appear in the balance equation of a scalar variable like a concentration. For this reason the mechanism discussed here gives no information on the selection of chirality, the second major form of asymmetry observed in nature.

Let us classify the patterns that can be generated by different chemical processes. Suppose first that the X_i tend asymptotically to a limit cycle. We know from Chapter 3 that this kind of behavior requires at least two coupled variables, and that the phase difference between any two of them will be constant for all times. It follows that the level crossings will at best generate patterns defined by the following symbolic dynamics if two variables manage to cross the assigned levels:

$$\cdots XYXYXY \cdots \tag{a}$$

or by

$$\cdots XYZXYZXYZ \cdots \tag{b}$$

if three variables manage to cross the assigned levels, and so on.

Sequence (a) above is reversible, that is, it reads the same in either direction, but sequence (b) describes an absolutely asymmetrical object. However, by introducing an abbreviated symbol (the analogue of a *codon*):

$$\alpha = \{XYZ\}$$

sequence (b) reads

$$\cdots \alpha\alpha\alpha \cdots \tag{c}$$

and in this form loses its space asymmetry.

Whether the structure is to be understood in terms of sequence (b) or (c), we can hardly speak of information, since the system is completely predictable. For instance, by applying Claude Shannon's definition of information (Eq. 4.43b), $I = -\sum_i P_i \ln P_i$, we would find I equal to zero.

Suppose now that the system runs on a chaotic attractor. As stressed in the preceding section and as we will see further in Chapter 5, under appropriate conditions the time evolution of an unstable dynamical system can be mapped to a stochastic process by means of an appropriate symbolic dynamics. In this sense, the transcript of chaotic dynamics according to the mechanism presented in this section may potentially give rise to structures that are:

1. Asymmetric because of irreversibility

2.   Information-rich, since the information theoretic entropy of a stochastic process is in general different from zero

Let us illustrate this conjecture with the results of numerical simulations using the Rössler model. Eqs. (3.30). For parameter values $a = 0.38$, $b = 0.3$, and $c = 4.5$, this model gives rise to a chaotic attractor, depicted in Figure 58. A reasonable set of threshold values turns out to be $L_x = L_y = L_z = 3.0$. Using the initial conditions $X_0 = Y_0 = Z_0 = 1.0$ we can use the level-crossing mechanism described above to generate a sequence of symbols of the form (printing starts only after a sufficiently long time for transients to die out has elapsed):

$$ZYXZXYXZXYXZYXZXYXZYXZYXZXZYXZYXZYXZXYXZYX \cdots$$

It can be checked that this sequence can be entirely reformulated by introducing the *hypersymbols*

$$\alpha = ZYX \qquad \beta = ZXYX \qquad \gamma = ZX$$

The result reads

$$\alpha\beta\beta\alpha\beta\alpha\alpha\gamma\alpha\alpha\beta\alpha \cdots$$

This interesting property is to be attributed to the deterministic origin of the mechanism that generates the sequences. It suggests the existence of strong correlations in the succession of the symbols X, Y, Z, that is to say, a high-order Markov process. To check this we evaluate the statistical properties of the sequences as follows. Eqs. (3.30) are integrated up to about 40,000 time units, which generates about 10,000 symbols and 3,000 hypersymbols. The results are recorded starting at $t = 300$, in order to have all transients die out. The numbers of observed singlets (X, Y, or Z), doublets (XX, XY, etc.), triplets, and so on, are counted. The conditional probabilities are then deduced through such relations as

$$P(B \mid A) = \frac{P(AB)}{P(A)}$$

$$P(C \mid AB) = \frac{P(ABC)}{P(AB)}$$

(4.44)

where A, B, C can be any of the X, Y, and Z*. The fact that doublets like XX never occur shows that the sequence is not completely random. Moreover, the

---

* *By increasing the integration time we can verify that the probability of finding a given symbol or a succession of symbols in a given order tends to a definite limit. It therefore seems that we are dealing with a stationary stochastic process.*

absence of the triplet YXY rules out the possibility that the sequence defines simple Markovian process since in that case

$$P(YXY) = P(Y)P(X|Y)P(Y|X)$$

all of which are different from zero. We therefore must look for a higher order Markov process in which the memory of an event extends several steps backward. As it turns out, good agreement with the numerically computed frequencies of septuplets of symbols is reached by assuming a fifth-order Markov chain:[*]

$$P(ABCDEFG) = P(ABCDEF)P(G|BCDEF)$$

$$= P(ABCDE)P(F|ABCDE)P(G|BCDEF)$$

where $P(G|BCDEF)$ represents the numerically computed conditional probability of the last symbol, given the quintuplet $BCDEF$.

A close inspection of the statistical properties of the sequences reveals some surprising features. For instance, of all possible 3^7 sequences of seven-symbol length that can be constructed from X, Y, and Z, only 21 are realized by the dynamics. Moreover, for about half of them the conditional probability of the last symbol, given the five preceding ones, turns out to be equal to one. Therefore, everything happens as if the system were endowed with a set of "grammatical" rules that are followed automatically as a result of the dynamics.

The above procedure can be repeated by changing the value of parameter a in the model, Eqs. (3.30), to 0.40. The results are qualitatively similar, including the possibility of expressing the sequences in terms of hypersymbols, except for a change in the numerical values of the probabilities of the different multiplets. Similarly, a small change in the threshold values leaves the main results unchanged. However, a large change in the value of the thresholds has drastic consequences. For instance, using the time average of X, Y, and Z as thresholds for these quantities, we find sequences that cannot be expressed entirely in terms of hypersymbols. Moreover, a Markov-chain model of order up to six is unable to fit the data adequately.

The information content of the sequence of X, Y, and Z can be computed from an extension of Shannon's definition, Eq. (4.43b), where the state i is now given by a quintuplet of symbols. It is found to be finite, but much less than the maximum corresponding to equiprobability. By producing these sequences as a natural outcome of its nonequilibrium dynamics, out of the much more numerous random sequences that can be envisaged, the system has thus become a generator of information. Notice, however, that each individual message remains unpredictable,

---

[*] *That the chain should be of at least fourth order could be inferred from the existence of the longest hypersymbol β and from the fact that Z is always preceded by X.*

as witnessed by the stochastic character of the sequence. On the other hand, the degree of unpredictability is tempered by the existence of correlations extending over the five nearest neighbors in the chain. To borrow a term used in communication theory, we can say that this enables the system to *compress* the information contained in the sequences of X, Y, Z by reproducing them more succinctly in terms of the hypersymbols α, β, γ. We thus find, in our simple model, one of the essential aspects of information pointed out in the begining of this section.

Let us summarize: At the outset we are confronted with the difficulty that the very existence of information automatically raises two issues that appear to be mutually contradictory. On the one side information must be associated with some kind of randomness—whence the numerous attempts found in the literature to relate it to "ignorance" and to "incomplete specification" of a system. But on the other side, the astronomically large number of random sequences of length $\rho \gg 1$ makes it extremely improbable to select, on a priori grounds, the particular class of sequences that is likely to play the major role in an observed phenomenon.

We have suggested a mechanism that allows us to come to terms with these restrictions: informationally meaningful structures can be generated from an underlying mechanism that is nonlinear, time-irreversible, and operating in the far-from-equilibrium chaotic region. In this way randomness and asymmetry—two prerequisites of information—are incorporated from the outset in the resulting structure. In addition, because these structures are the result of a mechanism (that ultimately has to do with the existence of molecules endowed with suitable catalytic properties), they automatically overcome the difficulty of the tremendous "thermodynamic improbability" that characterizes random sequences.

One of the most significant aspects of the concepts discussed in this and the preceding section is the emergence of a *new level of description* brought out by the underlying dynamics. Just as it becomes natural to describe the state of a nonequilibrium system in terms of the correlations between macroscopically separated elements rather than in terms of the intermolecular forces, and to describe the phenomenon of bifurcation in terms of the order parameter rather than in terms of the state variables initially present—so it now appears that in certain classes of stochastic dynamical systems it becomes in turn natural to introduce a still higher level of abstraction and speak of symbols and information. More than any other single characteristic among those enumerated and analyzed throughout these four chapters, it is this sweeping possibility that is to be regarded as the fingerprint of complexity.

4.9 ONCE AGAIN, ALGORITHMIC COMPLEXITY

At this stage the connection between physico-chemical complexity and the concept of *algorithmic complexity* introduced in Section 1.5 is worth considering again. As we have seen, algorithmic complexity is the length of the shortest description of a given (finite) sequence. In this sense, a completely random sequence generated by a noise process has the maximum possible algorithmic complexity, essentially equal to its length, whereas a sequence like XXX $\cdots$ has the minimum possible complexity since a single instruction suffices to reproduce it fully. Similarly, information is considered to be maximum in a specific random sequence since the realization of that particular sequence out of the enormous number of random sequences amounts to localizing the system very sharply in the state space. Information is zero in the minimum-complexity sequence since the outcome can only consist of sequences that are all alike.

We have already insisted that the complexity of natural objects lies somewhere between these two extremes, since in addition to randomness it also involves some large scale regularities. Erwin Schrödinger's celebrated aphorism that the DNA is an "aperiodic crystal" captures this duality in a particularly lucid and sharp manner.

Now, the self-organized states of matter allowed by nonequilibrium physics provide us with models of precisely this sort of complexity. Most important among these states, for our present purposes, is chaotic dynamics. Indeed, the instability of motion associated with chaos allows the system to explore its state space continuously, thereby creating information and complexity. On the other hand, being the result of a physical mechanism, these states are produced with probability one: the problem of selection of a particular sequence out of a very large number of a priori equally probable sequences simply does not arise. In a way, the dynamical system generating chaos acts as an efficient selector that rejects the vast majority of random sequences and keeps only those compatible with the underlying rate laws. Equally important perhaps, the irreversibility incorporated in these laws allows for the existence of attractors that have asymptotic stability and thus reproducibility. In concluding this chapter it is therefore legitimate to assert that we have finally identified a physical origin and some plausible mathematical architypes of complexity.

Toward a Unified Formulation of Complexity

*O*ne of the leading themes of this book is the dynamical basis of complexity. This involves both the deterministic theory of dynamical systems and the concepts of probability theory, especially Markov chains. We have emphasized the basic distinction that exists between conservative systems and dissipative systems, in Chapter 2. The variety of the tools needed to understand the dynamical basis of complexity reflects the wide range of mechanisms that lead to the birth of complexity, such as the onset of chaotic behavior in nonintegrable conservative systems, the symmetry breaking in dissipative systems, or the appearance of long-range correlations in nonequilibrium conditions.

What are the relations among these tools? Can we go from conservative systems to dissipative ones? What is the role of probability? We investigate these questions in this chapter. As dissipativity is closely related to the second law of thermodynamics, and entropy to probability (Section 4.3), we come back to one of the fundamental questions of physics: the meaning and the role of irreversibility.

This question has intrigued physicists since the very formulation of the second law in the middle of the nineteenth century. The second law was the outcome of centuries of observation, mainly in the fields of engineering and physical chemistry, and yet it has remained somewhat outside the mainstream of physics. It introduces a physical quantity, entropy, which endows time with a privileged direction: the "arrow of time," to use Arthur Eddington's expression. In contrast, since Newton, physics had set as its task the attainment of a basic level of reality that is timeless. The great conceptual revolutions brought about by relativity and quantum mechanics have not altered this program.

In the face of such a flagrant contradiction between the laws of conservative dynamical systems and the second law of thermodynamics, the majority of physicists have regarded the second law as arising from some approximation that we add to the laws of dynamics. Max Born, for instance, asserted that "irreversibility is a consequence of the explicit introduction of our ignorance into the fundamental [dynamical] laws."

However, it should be clear from the many examples examined here that irreversibility plays a fundamental role in nature. We simply cannot understand most of the phenomena around us, be it the growth of crystals, the regulation of biological systems, or the history of climate, without appealing to dissipative systems. In quite different fields of physics, the role attributed to irreversibility is steadily growing. Who would have expected that most or perhaps all elementary particles are unstable, that irreversible processes played such an essential role in the early stages of cosmological evolution? It therefore becomes impossible to simply dismiss irreversibility as contrary to the traditionally accepted view of physics. Before we proceed to the dynamical formulation of the second law, it seems appropriate to summarize the basic properties of conservative and dissipative dynamical systems as discussed in the preceding chapters.

5.1 GENERAL PROPERTIES OF CONSERVATIVE DYNAMICAL SYSTEMS

The evolution of a conservative dynamical system corresponds to a point transformation (which we shall call S_t) in an appropriate phase space Γ:

$$\omega_t = S_t \omega_0 \tag{5.1a}$$

where ω_t and ω_0 are the phase space points at time t and time 0, respectively. A fundamental property of these systems as derived from the Liouville theorem is that the flow in phase space is that of an incompressible fluid. In other words, volume is preserved in phase space (see Section 3.2). The dynamics may also be expressed in terms of a unitary operator U_t acting on the distribution function (see Section 4.3):

$$\rho_t(\omega) = U_t\rho_0(\omega) \tag{5.1b}$$

$\rho_t(\omega)$ is the distribution function at time t, and $\rho_0(\omega)$ its initial value.* As we have noticed [see Eq. (4.22b)], the U_t form a dynamical group

$$U_t U_s = U_{t+s} \ (t, \ s \text{ positive or negative}) \tag{5.2}$$

This formal description of dynamical systems includes classes of systems having widely different behavior. In classical physics, integrable systems such as the frictionless pendulum or the two-body problem were at the center of interest. We know now that this is an exceptionally simple class, one that gives an inadequate idea of the true complexity of dynamical systems. The discovery of the variety of dynamical behaviors of nonintegrable systems therefore marks one of the major breakthroughs of contemporary physics and mathematics (see Section 3.8).

What would be the properties of a world dominated by conservative dynamical systems?[†] Perhaps the basic points which we have mentioned repeatedly, is that such a world would be unstable, as the property of asymptotic stability which permits to a system to forget accidental perturbations does not apply to conservative dynamical systems (see Section 2.6). At best we would have "orbital stability" around elliptic points, a state in which perturbations shift the system from one orbit to the other. The situation would be worse around hyperbolic points, for there perturbations lead to instability. Moreover, there is in general an abundance of elliptic and hyperbolic points. As a result, we must expect an extreme sensitivity to initial conditions. We have already seen an example of this kind of behavior in the horseshoe, and we shall introduce as an even simpler illustration of orbital randomness, the Baker transformation, in Section 5.5.

Motion involving hard spheres gives another example of unstable systems. A much-studied system is based on the Lorentz model, named for the Dutch physicist Hendrik Antoon Lorentz. In this model, one considers the motion of a small sphere rebounding on a collection of randomly distributed large spheres that are fixed in space. This is represented in Figure 82. The trajectory of the

---

* *Since the motion can be expressed in terms of trajectories as described by Eq. (5.1a), Eq. (5.1b) is equivalent to $\rho_t(\omega_t) = \rho_0(\omega_0)$, or $\rho_t(\omega) = \rho_0(S_{-t}\omega)$.*
[†] *In the sense that macroscopic observables would follow a time-reversible, conservative dynamics.*

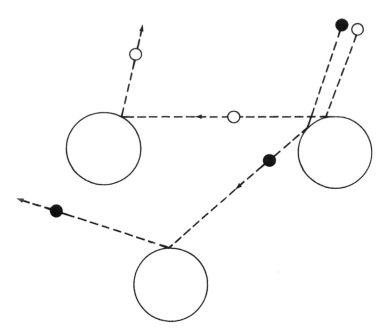

Figure 82 Schematic representation of the instability of the trajectory of a small sphere rebounding on large spheres. The least imprecision about the initial position of the small sphere makes it impossible to predict which large sphere it will hit after a few collisions.

small mobile sphere is well defined. Whenever we introduce the smallest uncertainty in the initial conditions, this uncertainty is increasingly amplified through successive collisions. Note that chemical reactions may often be considered as the outcome of collisions between hard spheres in which the kinetic energy of the relative motion, the so-called activation energy, is beyond some threshold. Again, such motions are expected to be strongly unstable.

While the volume (the measure) in phase space is always conserved during the evolution, its shape is highly deformed as a result of instability. As we shall see in Section 5.5, the initial volume may be fragmented in smaller and smaller pieces as time goes on. It is this deformation or fragmentation that gives the appearance of an approach to equilibrium, in which all points would be uniformly distributed in the accessible region of phase space. For nearly three centuries after Newton, classical dynamics appeared as a closed science, enabling us to compute any evolution from first principles and well-defined initial data. We now see that this holds only for limited classes of dynamical systems. For sufficiently unstable dynamical systems each region of phase space, whatever its size, contains diverging trajectories. To be able to speak of a well-defined single trajectory, we would need initial data involving a strictly infinite precision. In other words, we would need an infinite amount of information (associated with the infinite number of digits required to describe the initial data). It is the elimination of this infinite information which leads, as we shall see, to irreversibility.*

In the world of unstable dynamical systems we can only look through a "window" on the outside world. We witness here the breakdown of the ideal of complete knowledge that has haunted Western science for three centuries.

5.2 GENERAL PROPERTIES OF DISSIPATIVE DYNAMICAL SYSTEMS

Dissipative dynamical systems present an extraordinary variety. We have looked at many examples pertaining to chemistry, fluid dynamics, and other fields of the natural sciences. We know that, in contrast to conservative systems, dissipative systems do not conserve the measure in phase space (see Section 3.2). Moreover, they have in common the crucial element that they are not invariant in respect to time reversal. It follows that some privileged situation will be reached asymptotically, as $t \to +\infty$. If we do not impose any constraints, this will be the equilibrium situation. For example, the concentration or the temperature, as described by Eqs. (2.7b), (2.7c) will tend to uniformity; similarly, the concentrations corresponding to a reactive mixture will tend to values according to the law of mass action. If nonequilibrium constraints are imposed, the evolution will tend to some attracting set, be it a point attractor, a limit cycle, or a "strange" attractor as in Section 3.10. What is essential is that in all dissipative systems there exists a preferential direction of time. We could imagine a world in which some biological systems age while others become younger; a world in which some dissipative systems tend to equilibrium for $t \to +\infty$ while others do so for $t \to -\infty$. But this obviously is not our world, in which a universal time symmetry breaking is present. Fortunately so, as noted by Norbert Wiener: in any world within which we can communicate, the direction of time is bound to be uniform. In contrast with the orbital randomness of conservative dynamical systems, we find asymptotic stability here. This makes possible the appearance of delicately balanced processes, such as we see for instance in biology. However, randomness may still persist at the level of attractors.

In section 3.10 we studied various models in which chaotic behavior can be observed. However, the Lyapounov exponents, which according to Appendix 1 characterize the instability of motion, are now determined by macroscopic quantities such as chemical rates and diffusion coefficients, not by the microscopic dynamics.

---

* *We limit ourselves to classical mechanics here; in quantum theory the decay of unstable particles is an additional source of irreversibility.*

Not only is dissipative behavior associated with a universal time symmetry breaking; in addition we can associate dissipative processes with a Lyapounov function such as entropy that changes monotonically in time. Indeed, all dissipative processes we have been studying lead to a positive entropy production (Section 2.5). The very existence of dissipative dynamical systems is a manifestation of the second law of thermodynamics.

5.3 THE SEARCH FOR UNIFICATION

There is still another point of view that permits us to contrast conservative with dissipative dynamical systems. It seems impossible to find a more fundamental level of description out of which conservative dynamical systems would emerge as the result of approximations. Of course, if we introduce quantum theory, classical trajectories become averages over wave packets. This however does not lead to a new point of view, since the Schrödinger equation—the basic equation of quantum theory—is deterministic and time reversible, just like the equations of classical dynamics*.

The situation is quite different for dissipative systems. As an example let us consider chemical kinetics. The equations encountered therein do not correspond to a "fundamental level" since, as we have seen, they result from approximations in which *fluctuations* are neglected [see especially Eq. (4.15)]. We also observe *correlations* that reflect the coherent behavior arising in nonequilibrium situations. The fundamental level of dissipative systems has therefore to be associated with a *probabilistic approach* such as the one provided by the theory of Markov chains studied in Chapter 4.

Can this markovian formulation be extended from the state space of macroscopic observables, to which Chapter 4 was limited, to the full phase space Γ in which so far only reversible descriptions, Eqs. (5.1), were available? We shall try this by introducing in Γ a distribution function $\tilde{\rho}(\omega)$ which, in contrast to classical mechanics, now evolves according to a Markov process. In analogy with Eq. (5.1b) we may write for $\tilde{\rho}_t$ at time t

$$\tilde{\rho}_t(\omega) = W_t \tilde{\rho}_0(\omega) \tag{5.3}$$

it being understood that $\tilde{\rho}_t(\omega)$ is the solution of a master equation describing the Markov process. As we have seen, the reversible description based on Liouville's equation satisfies the group property, Eq. (5.2). In contrast, W_t satisfies the semigroup condition

$$W_t W_s = W_{t+s} \qquad (t, s \geqslant 0) \tag{5.4}$$

---

* Similarly, it would not help to replace particle dynamics by field equations such as are used in electrodynamics or relativity. Again, the basic field equations are deterministic and time reversible.

The Markov process, Eq. (5.3), with the condition of Eq. (5.4), drives the system to equilibrium, or to some nonequilibrium state characterized by suitable attractors when constraints are introduced.

In summary, in conservative dynamics there seems to be no place for genuine probabilistic evolution processes, but such processes form the basis of the description of dissipative systems. Therefore, the elucidation of the relation between conservative and dissipative dynamical systems necessarily involves a clarification of the relation between deterministic dynamics and probabilities. Because of the close relation that exists between entropy and probability, once this is clarified the relation between dynamics and the second law will also be made clear. As the basic laws of reversible conservative dynamics may be expressed through Eqs. (5.1), while genuine probabilistic processes satisfy the relations in Eqs. (5.3) and (5.4), we have to discuss the way in which all of these expressions might be connected.

5.4 PROBABILITY AND DYNAMICS

The basic quantity in the theory of Markov processes is the transition probability. As we want to set up a genuine probabilistic description in phase space Γ, we consider the transition probability $P_t(\omega|\Delta)$ from point ω to some domain Δ at time t, which is a nonnegative number between zero and one. Here we come to a basic difference between markovian description and description in terms of trajectories.

Suppose that a trajectory goes from ω to ω_t, as in Figure 83. Then, for a system following the trajectory we obviously have

$$
\begin{aligned}
P_t(\omega|\Delta) &= 1 && \text{if } \omega_t \text{ is in } \Delta \\
&= 0 && \text{if } \omega_t \text{ is not in } \Delta
\end{aligned}
\tag{5.5}
$$

In contrast, in a proper Markov process at least some of the transition probabilities are neither zero nor one.

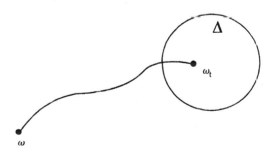

Figure 83 Transition probability $P_t(\omega|\Delta) = 1$, characteristic of a deterministic description based on trajectories.

We have no reason to assume that the dynamical description, Eqs. (5.1), is wrong. However it may be incomplete, for it does not incorporate the element of nonlocality that is obviously necessary to bridge the gap between conservative and dissipative systems by leading from the degenerate transition probabilities, Eqs. (5.5), to proper markovian ones. In view of this situation it is tempting to connect the descriptions of Eqs. (5.1) and (5.3) through a transformation,

$$\tilde{\rho} = \wedge \rho \tag{5.6}$$

where the transformation operator $\wedge$ acting on ρ should incorporate this missing nonlocality, and must of course depend on the laws of dynamics. Once we succeed in determining $\wedge$, Eq. (5.6) will provide a link between the distribution function ρ evolving according to reversible dynamics and the distribution function $\tilde{\rho}$ evolving according to a Markov process.

The relation in Eq. (5.6) implies a remarkable interwining between U_t and W_t. Indeed, inserting Eqs. (5.1b) and (5.3) into Eq. (5.6) we get

$$\wedge U_t \rho_0 = W_t \wedge \rho_0 \tag{5.7}$$

Since this is true whatever ρ_0, we conclude

$$\wedge U_t = W_t \wedge \tag{5.8}$$

If $\wedge$ admits an inverse, Eq. (5.8) may be written

$$W_t = \wedge U_t \wedge^{-1}$$
$$U_t = \wedge^{-1} W_t \wedge \tag{5.9}$$

In this view, probability theory is an "image" of dynamics mediated by the transformation $\wedge$. Inversely, the dynamics underlying a Markov process is recovered whenever we are able to identify $\wedge$. The central problem we face in building a bridge between deterministic dynamics and probabilities therefore lies in the construction of $\wedge$. Let us turn now to a class of conservative dynamical systems for which this construction can be explicitly achieved.

5.5 THE BAKER TRANSFORMATION

We consider the class of highly unstable dynamical systems called K-flows (K for Kolmogorov). In Section 3.9 we studied the horseshoe example. Here we study another celebrated example, the Baker transformation, illustrated in Figure 84. On

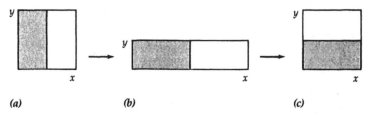

Figure 84 The Baker transformation. The unit square, (a), is flattened into a $\frac{1}{2} \times 2$ rectangle, (b). The right half of the rectangle is then superposed on the left part to form a new square, (c).

the phase space $0 \leqslant x \leqslant 1, 0 \leqslant y \leqslant 1$ (the unit square), we define the mapping

$$x' = 2x, \qquad y' = \frac{y}{2} \qquad \text{if } 0 \leqslant x < \tfrac{1}{2}$$

$$x' = 2x - 1, \quad y' = \frac{y + 1}{2} \qquad \text{if } \tfrac{1}{2} \leqslant x \leqslant 1$$

(5.10)

This transformation corresponds to a measure-preserving discrete evolution process occurring at regular times, for example every second (the measure here is the surface area.) The transformation therefore defines a conservative dynamical system. It is also obviously invertible, since the unit square is mapped onto itself (contrary to what happens in the horseshoe transformation).

The time reversibility of the dynamics is somewhat subtler to establish. Let us solve Eqs. (5.10) for x and y:

$$x = \frac{x'}{2}, \qquad y = 2y' \qquad 0 \leqslant y < \tfrac{1}{2}$$

$$x = \frac{x' + 1}{2}, \quad y = 2y' - 1 \qquad \tfrac{1}{2} \leqslant y \leqslant 1$$

(5.11)

We see that Eqs. (5.11) are identical to the time-reversed version of Eqs. (5.10) provided that x and y are interchanged. This interchange is the analogue of the inversion of momenta (keeping spatial coordinates unchanged) when reversing time in the equations of hamiltonian mechanics (see Section 2.1).

In Eqs. (5.10) we observe that one direction, x, is expanding and the other, y is contracting. After n consecutive transformations, the distance between two points on the x axis will be multiplied by a factor $2^n = e^{n \ln 2}$. According to the definitions given in Appendix 1 we therefore have a positive Lyapounov exponent,

$$\sigma_1 = \ln 2 \tag{5.12}$$

This establishes the chaotic character of the dynamics. Naturally, as the system is conservative, the second Lyapounov exponent is negative, $\sigma_2 = -\ln 2$ [see Eq. (A1.27)].

It can also be seen, just by repeating the process indicated in Figure 84, that as time goes on each finite subregion will be partitioned into finer and finer strips until it becomes distributed practically uniformly over the unit square. This is represented in Figure 85.

Much insight in the mechanism of the Baker transformation is gained by representing it as a process known as the *Bernoulli shift*. To do so, we associate with each point (x, y) of the unit square the doubly infinite sequence of numbers $\{u_n\}$ defined by the binary representation:

$$x = \sum_{n=-\infty}^{0} \frac{u_{n\cdot}}{2^{-n+1}},$$

$$y = \sum_{n=1}^{\infty} \frac{u_n}{2^n}$$

(5.13)

where each u_n can take the values 0 or 1. Inserting this representation in Eqs. (5.10) we obtain:

$$\frac{u_0{}'}{2} + \frac{u'_{-1}}{2^2} + \cdots = x' = \frac{u_{-1}}{2} + \frac{u_{-2}}{2^2} + \cdots$$

$$\frac{u_1{}'}{2} + \frac{u_2{}'}{2^2} + \cdots = y' = \frac{u_0}{2} + \frac{u_1}{2^2} + \cdots$$

The time evolution induced by the Baker transformation therefore corresponds to the shift

$$u_n{}' = u_{n-1}$$

(5.14)

referred to as the Bernoulli shift. We see that, as expected, the information contained in the initial conditions includes all past and future history of the system.

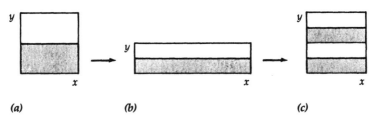

(a)   (b)   (c)

Figure 85 Successive iterations of the Baker transformation lead to fragmentation of the shaded and unshaded areas into an increasing number of disconnected regions.

Notice that the digit u_0 determines whether the representative phase space point is in the left half ($u_0 = 0$) or in the right half ($u_0 = 1$) of the unit square. Since the digits u_0, u_{-1}, u_{-2}, ... can be determined by tossing a coin, the time iterates of u_0, $u_0{}' = u_{-1}$, $u_0{}'' = u_{-2}$ will have the same random properties as coin tossing. This shows that the event of the system being left or right can be considered to be a Bernoulli process, one of the most familiar examples of random processes in probability theory.

The Baker transformation also shares the important property of *recurrence* of states. Consider a point (x, y) for which the sequence $\{u_n\}$ in the binary digit representation, Eqs. (5.13), is finite or infinitely periodic. x and y are rational in this case. Since all u_n are shifted in the same way by the dynamics, all states of this kind will recycle exactly after a certain period of time. More importantly, the same holds true, in a weaker sense to be specified presently, for almost all other states (except for a set of measure zero). To see this we consider the binary representation of an irrational point (x, y), which contains an infinity of nontrivial, nonrepeating digits. It can be shown that almost all irrationals (except for a set of measure zero) contain any finite sequence of digits infinitely often. Thus, a given sequence of, say, $2m$ digits around position 0, which determines the state up to an error of 2^{-m}, will reappear infinitely often under the effect of the shift. Since m can be chosen as large as desired (although finite), this means that almost every state will return an infinite number of times arbitrarily close to any point, including of course the initial position. In other words, most of the trajectories will be *ergodic* and traverse the whole phase space. This is the famous Poincaré recurrence theorem, which originally was proved for a very general class of conservative dynamical systems. For a long time, recurrence, together with time reversibility, was advanced as an essential argument against the existence of genuinely dissipative processes. As we shall see shortly, this view can no longer be sustained.

In summary, the dynamics described by the Baker transformation is conservative, invertible, time reversible, recurrent, and chaotic. The manifestation of all these properties makes the Baker transformation an especially useful example, since these same properties characterize real-world dynamical systems showing complex behavior. It is striking that despite reversibility and recurrence, the property of chaoticity will allow us to establish genuine irreversibility by setting up, without introducing any approximation, a more adequate description.

The dynamics of conservative systems involves laws of motion and initial conditions. Here the laws of motion are simple, but let us analyze the concept of initial data more closely. The initial data of a single trajectory correspond to an infinite set $\{u_n\}$ ($n = -\infty$ to $+\infty$). But in the real world we look only through a finite window. In the present case this means that we control an arbitrary but limited number of digits u_n. Suppose that this window corresponds to

$$u_{-3}u_{-2}u_{-1}u_0 \cdot u_1 u_2 u_3 \tag{5.15}$$

all other digits being unknown (the dot indicates the separation between x and

y digits). The Bernoulli shift implies that at the next step, Eq. (5.15) is replaced by

$$u_{-4}u_{-3}u_{-2}u_{-1} \cdot u_0u_1u_2 \tag{5.16}$$

which contains the unknown digit u_{-4}. More precisely, owing to the existence of a positive Lyapounov exponent, we need to know the initial position with an accuracy of $N + n$ digits in order to be able to determine the position with an accuracy of N digits after n iterations.

In statistical mechanics, the traditional way to cope with this situation would be to introduce a coarse-grained probability distribution in the unit square. Such a distribution is not defined on single points, but on regions, as proposed originally by Paul and Tatiana Ehrenfest. However, two points on an expanding manifold, even if not distinguishable by measurements of a given finite precision at time 0, will be separated later and could then be distinguished. The traditional coarse graining therefore does not commute with the dynamical evolution. This is the reason we need a more sophisticated method, which we shall develop in the next two sections.

5.6 MANIFOLDS WITH BROKEN TIME SYMMETRY

In spite of the fact that K-flows are time reversible, they contain objects displaying broken time symmetry. These objects are of lower dimensionality than the full phase space, and in the case of the Baker transformation, they are "contracting fibers" and "dilating fibers," as represented in Figure 86. Contracting

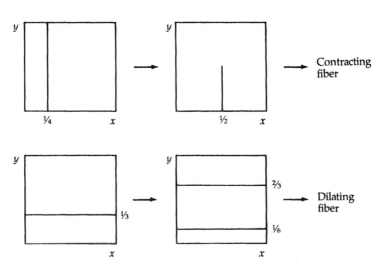

Figure 86 Contracting and dilating fibers in the Baker transformation.

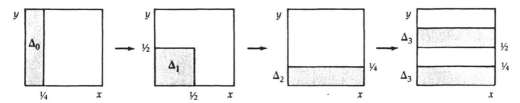

Figure 87 Inadequacy of the Liouville theorem when observation involves only a finite window (here $\varepsilon = \frac{1}{4}$) on the contracting fibers. The measure of initial volume Δ_0 remains invariant for a few iterations, but eventually increases until it reaches the total measure of phase space.

fibers are vertical lines that contract by a factor of 2 at each Baker transformation; dilating fibers are horizontal lines that expand by the same factor. Each phase space point is the intersection of a contracting and a dilating fiber. Contracting and dilating manifolds are the time inverses of each other and are obviously time-asymmetric objects. All points on a contracting manifold tend to the same future. In contrast, the points on a dilating fiber will densely cover the phase space at $t \to +\infty$. In a probabilistic language, this corresponds to the approach to uniform equilibrium distribution in phase space.

Whatever our precision in determining a state might be, there will always be a characteristic distance ε below which we cannot distinguish points on the contracting fibers. As points on such fibers belong to converging trajectories, *all trajectories starting inside ε must be considered indiscernible in the future.* This immediately makes the Liouville theorem expressing the conservation of measure in phase space irrelevant. A simple argument is the following. Let us consider a dynamical system, and suppose that the distribution function ρ is confined for $t = 0$ to an area Δ_0. As time goes on, the dilating dimension will increase exponentially, while the contracting dimension will shrink to ε *and must be considered constant thereafter.* (The procedure is shown in Figure 87.) As a result, the measure Δ will increase from then on, until it reaches the measure of the total phase space. Note that this conclusion does not result from an approximation. It is the outcome of the basic fact that measurements (or, more generally, effects of physical interactions) refer to *regions* of finite measure in phase space, not to mathematical points. It was of course always understood that exact initial conditions correspond to an idealization. What is new is that giving up this idealization leads to so momentous a consequence in highly unstable systems such as K-flows.

5.7 THE SYMMETRY-BREAKING TRANSFORMATION $\wedge$

Let us now indicate how the ideas put forth in the preceding section lead to the construction of the symmetry-breaking transformation $\wedge$, Eq. (5.6), which links dynamical and probabilistic processes.

To this end, it is useful to represent the state of the unstable dynamical system under consideration in terms of objects referred to as *partitions*, $\{\chi_n\}$. In the case of the Baker transformation, it is very easy to construct such partitions. Indeed, let χ_0 be the function that assumes the values -1 on the left half of the unit square, and $+1$ on the right half. We define χ_n as being the result of n successive transformations (n being a positive or negative integer) acting on χ_0:

$$\chi_n = U^n \chi_0 \tag{5.17}$$

where U denotes the evolution operator for distribution functions, Eq. (5.1b). A few of these partitions are represented in Figure 88. Note that χ_0 is entirely determined by the digit u_0 in the representation of Eqs. (5.13) whereas in Eq. (5.17) χ_n is determined by the digit u_n.

Let us now see how we may describe the dynamics in terms of partitions. To do so, it is useful to characterize the partitions χ_n by their "age," n. This concept, which in view of Eq. (5.17) is quite natural, can be formalized and extended by means of an important result of Baidyanath Misra, establishing the existence of an operator T that admits χ_n and n respectively as eigenfunctions and eigenvalues, $T\chi_n = n\chi_n$. Actually, each eigenvalue n of this operator displays an infinite degeneracy, since besides χ_n, T possesses all products

$$\phi_n = \chi_{j_1}\chi_{j_2}\cdots\chi_n \qquad (j_1 < j_2 < \cdots < n)$$

as eigenfunctions. We therefore write

$$T\phi_n = n\phi_n \qquad (-\infty \leqslant n \leqslant \infty) \tag{5.18a}$$

To complete the definition of T we require that $\phi \equiv 1$ be an additional eigenfunction with eigenvalue 1,

$$T \cdot 1 = 1 \tag{5.18b}$$

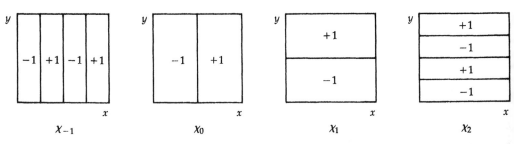

Figure 88 Representation of the dynamics of the Baker transformation in terms of partitions.

and that T be a hermitean linear operator. The functions ϕ_n together with $\phi = 1$ can be shown to form a complete set.

Notice that the effect of T on a partition χ_n is different from that of the dynamical operator U, since according to Eq. (5.17),

$$U\chi_n = \chi_{n+1} \tag{5.19}$$

As a consequence of this difference, T and U are noncommuting operators. From Eqs. (5.18) and (5.19)

$$TU\chi_n = T\chi_{n+1} = (n + 1)\chi_{n+1}$$

On the other hand

$$UT\chi_n = Un\chi_n = n\chi_{n+1}$$

Thus,

$$(TU - UT)\chi_n = \chi_{n+1} \neq 0$$

It is remarkable that in this way classical dynamics leads to a noncommutative algebra not unlike that of quantum mechanics.

We now introduce the requirement that the symmetry-breaking transformation Λ be a hermitean linear operator that is a nonnegative decreasing function of T in the sense that

$$\Lambda(T)\phi_n = \lambda_n\phi_n$$
$$\Lambda(T)1 = 1 \tag{5.20}$$

with

$$1 \geqslant \cdots \geqslant \lambda_n \geqslant \lambda_{n+1} \geqslant \cdots \geqslant 0$$

and

$$\lambda_n \rightarrow \begin{cases} 1 & \text{for } n \rightarrow -\infty \\ 0 & \text{for } n \rightarrow +\infty \end{cases}$$

Since the partitions ϕ_n together with the uniform distribution 1 form a complete set, we can expand any distribution function ρ as

$$\rho = 1 + \sum_{n=-\infty}^{+\infty} c_n\phi_n \tag{5.21}$$

In writing this relation we have suppressed, for the sake of simplicity, an additional index referring to the degeneracy of the eigenvalues of T. Using Eqs. (5.20) and $\tilde{\rho} = \wedge \rho$ if follows that

$$\tilde{\rho} = 1 + \sum_{n=-\infty}^{+\infty} c_n \lambda_n \phi_n \qquad (5.22)$$

Note that at a given point of phase space the passage from ρ to $\tilde{\rho}$ involves all points of phase space. This establishes the expected nonlocality of the operator $\wedge$.

All states ϕ_n enter into ρ as such, their effect being weighted only through the factors c_n prescribed by the initial condition. In this respect no distinction is made between past ($n < 0$) and future ($n > 0$). In contrast, in $\tilde{\rho}$ for given initial conditions the contribution of the ϕ_n's pertaining to the future is damped ($\lambda_n \to 0$ for $n \to \infty$). Now, remember that partition χ_n is determined by the digit u_n. In addition the digits u_n with $n > 0$ specify entirely the contracting fibers [See Eqs. (5.13)]. Therefore, the damping $\lambda_n \to 0$ expresses the fact that we lose information concerning the localization on these fibers. This corresponds exactly to the idea of the indiscernibility of neighboring points. For example, with $\lambda_n = 1$ for $n \leqslant 2$ and $\lambda_n = 0$ for $n > 2$, the natural value of the distance ε introduced in Section 5.6 is $\varepsilon = \frac{1}{8}$. We have presented in Section 5.6 an intuitive argument showing why indiscernibility makes the Liouville theorem useless.

Let us verify that for the new quantity $\tilde{\rho}$ we can now define a function that evolves monotonically, in conformity with the second law of thermodynamics (see Section 4.3). We consider the so-called $\mathscr{H}$-function

$$\mathscr{H}_{\tilde{\rho}} = \int \tilde{\rho}^2 \, d\omega \qquad (5.23)$$

(other convex functions, such as $\int \tilde{\rho} \ln \tilde{\rho} \, d\omega$ could be introduced with similar results). We suppose that at time $t = 0$, ρ and $\tilde{\rho}$ are given by Eqs. (5.21) and (5.22), respectively. Then, using Eq. (5.19) and the obvious fact $U \cdot 1 = 1$,

$$\rho_1 = U\rho_0 = 1 + \sum_n c_n \chi_{n+1} \qquad (5.24)$$

$$\tilde{\rho}_1 = \wedge \rho_1 = 1 + \sum_n c_n \lambda_{n+1} \chi_{n+1} \qquad (5.25)$$

As a result we obtain from Eq. (5.23)

$$\mathscr{H}_{\tilde{\rho}_1} - \mathscr{H}_{\tilde{\rho}_0} = \sum_n (\lambda_{n+1}^2 - \lambda_n^2) c_n^2 \leqslant 0 \qquad (5.26)$$

that is, $\mathscr{H}_{\tilde{\rho}}$ decreases monotonically with time. In contrast, the quantity $\mathscr{H}_\rho = \int \rho^2 \, d\omega$ expressed in terms of the original distribution function ρ remains constant. The result just obtained justifies the monotonicity assumption made in introducing the operator $\wedge$, see Eqs. (5.20). As a matter of fact, monotonicity of the λ_n's is both necessary and sufficient for the validity of Eq. (5.26).

In summary, we have arrived at our goal of constructing the transformation $\wedge$ and obtaining a microscopic formulation of the second law of thermodynamics. This highlights the importance of the distribution $\tilde{\rho}$. As a matter of fact, it is understood that not only entropy (via $\mathcal{H}_{\tilde{\rho}}$) but also all other observables are to be expressed in terms of $\tilde{\rho}$.

So far we have only assumed that the transformation operator $\wedge(T)$ is a decreasing but otherwise undetermined function of T. We cannot go into a detailed discussion on the form of $\wedge(T)$ here, nor alternatively of λ_n as a function of n. As the only intrinsic parameter of our model is the Lyapounov exponent, Eq. (5.12), it is reasonable to expect that at least for n positive and large

$$\lambda_n \sim e^{-n/\tau} \tag{5.27}$$

where $1/\tau = \ln 2$ is the Lyapounov exponent. Whatever the detailed form of λ_n, the fundamental properties of $\wedge$ are nonlocality and broken time symmetry. These are the properties that lead from unstable conservative dynamical systems to dissipative systems.

5.8 GIBBS ENSEMBLES AND BOLTZMANN ENSEMBLES

Classical dynamics leads to a point transformation. In the language of Gibbs ensembles, for a point in phase space there corresponds a Gibbs ensemble represented by a δ function. The motion $\omega_0 \rightarrow \omega_t$ transforms one δ function into another:

$$\delta_{\omega_t} = U_t \, \delta_{\omega_0} \tag{5.28}$$

where $\delta_{\omega_0} = \delta(\omega - \omega_0)$, etc. The δ function representation is therefore invariant under the action of the unitary operator U_t. Let us now apply the transformation $\wedge$ to δ_{ω_0}:

$$\tilde{\delta}_{\omega_0} = \wedge \, \delta_{\omega_0} \tag{5.29}$$

In general, $\tilde{\delta}_{\omega_0}$ is not strictly localized. However, due to the monotonicity of the λ_n's it is concentrated very strongly on the contracting fiber through ω_0, along which it presents a more or less broad distribution with a maximum around ω_0. Figure 89 depicts an example in which $\tilde{\delta}_{\omega_0}$ is entirely concentrated on the contracting fiber. The time evolution of $\tilde{\delta}_{\omega_0}$ is of course given by Eq. (5.3), $\tilde{\delta}_{\omega_t} = W_t \, \tilde{\delta}_{\omega_0}$. In the case of Figure 89, it can be shown that the action of W_t amounts to a rigid displacement of the support of $\tilde{\delta}$.

In the traditional representation of dynamics, the basic "units" are points; now they are the nonlocal entities $\tilde{\delta}_{\omega_0}$ that are essentially associated with segments

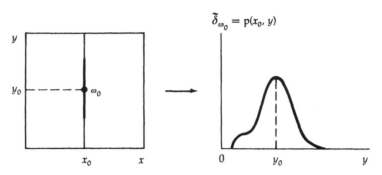

Figure 89 Typical form of a density $\tilde{\delta}_{\omega_0}$ on a contracting fiber.

of the contracting fibers. Instead of single trajectories represented by δ functions, we are now dealing with *bundles of trajectories* represented by $\tilde{\delta}$ functions.

The Gibbs ensembles may be considered as superpositions of phase space points, as

$$\rho(\omega) = \int_\Gamma \rho(\omega_0)\, \delta(\omega - \omega_0)\, d\omega_0 \tag{5.30}$$

This construction has to be amended in the case of unstable dynamical systems by going from ρ to $\tilde{\rho}$. By applying $\wedge$ on both sides of Eq. (5.30) we see that the ensembles $\tilde{\rho}$ are superpositions of $\tilde{\delta}$ functions:

$$\tilde{\rho}(\omega) = \wedge \rho(\omega)$$
$$= \int_\Gamma \rho(\omega_0)\, \tilde{\delta}(\omega - \omega_0)\, d\omega_0 \tag{5.31}$$

These ensembles may be called *Boltzmann ensembles* as they evolve according to a Markov semigroup, and lead to a monotonic increase of entropy. The transition from Boltzmann to Gibbs ensembles corresponds to the transition from arbitrary finite to infinite precision. *The Gibbs ensemble therefore constitutes a singular, nonphysical limit of the Boltzmann ensemble.* Let us emphasize once again how much this differs from the attitude in which irreversibility is attributed to an approximation introduced in the description of reversible processes.

5.9 KINETIC THEORY

We have so far devoted this chapter to a specific class of unstable dynamical systems, the K-flows, and especially to a simple example, the Baker transformation. A number of more realistic situations, such as brownian motion, or the Lorentz model (Section 5.1) can be studied using the methods developed.

We devote this section to another question: why is irreversibility so common in large systems such as gases or liquids? What is the origin of instability in such systems? This is far from being a trivial question. Hardsphere models such as the Lorentz model are not hamiltonian systems, as the potential energy of two hard spheres is either infinite or zero, not a continuous function of distance. Therefore, the Hamilton equations of motion, Eqs. (2.5b), do not apply. In contrast, in real multibody systems we have smooth interactions and therefore a hamiltonian containing both kinetic and potential energy. As we want to show now, we can begin to understand why such large systems can be very close to K-flows.

Let us go back to Chapter 3. In Section 3.8 we emphasized the role of the resonance conditions. In the low-dimensional systems considered there, on most tori the frequencies are incommensurate (nonresonant tori); this leads to quasi-periodic motion surrounding elliptic points. However, we also have a dense distribution of hyperbolic points arising from the destruction of resonant tori. We want to show now that in the limit of large systems the situation is likely to change drastically, as the resonance condition should be satisfied almost everywhere.

Let us start by observing that in such systems the dynamics can be viewed as a free-flight motion interrupted by successive encounters between particles. This interplay is seen most clearly in the neighborhood of the free particle limit by a formal perturbative solution of the Liouville equation, in which only the dominant terms are retained. By also invoking the assumption that the initial correlations between particles are short range, we arrive at an equation referred to as a *kinetic equation* that governs the relaxation of the system to equilibrium. This equation involves an object that plays a central role in modern kinetic theory, the *collision operator*. In the dominant order of perturbation theory this operator contains a multiplicative contribution of the form

$$\left(\frac{2\pi}{L}\right)^3 \sum_l |V_l|^2 l^2 \, \delta[l \cdot (p_j - p_k)] \tag{5.32}$$

in which p_j, p_k are the momenta of particles j and k, and δ is the Dirac delta function. V_l is the Fourier transform of the interaction potential V of particles j and k,

$$V(|x_j - x_k|) = \left(\frac{2\pi}{L}\right)^3 \sum_l V_l e^{il \cdot (x_j - x_k)} \tag{5.33}$$

$$l = \frac{2\pi}{L} n$$

n being a vector with integer components. In writing this last relation we assume periodic boundary conditions and that V depends only on the distance $|x_j - x_k|$ between the two particles. The factor $(2\pi/L)^3$, L^3 being the volume of the system, ensures the transition from Fourier series to Fourier integral in the limit of a large system, $L \rightarrow \infty$.

The validity of kinetic theory implies that the expression displayed in Eq. (5.32) exists and is finite in a whole region of phase space. Let us analyze the consequences of this property for the dynamics. Owing to the presence of the delta function, we obtain nonvanishing contributions in Eq. (5.32) for those l's (with $|l| \neq 0$) for which

$$l \cdot (p_j - p_k) = 0 \qquad (5.34)$$

What is the meaning of this relation? From the second of Eqs. (5.33) l is equal (up to a multiplicative factor) to a vector with integer components. Moreover, in agreement with the perturbative approach p_j and p_k are associated with free particle motion and hence are related to the inverse of the time needed by the particles to cross the characteristic dimension L of the system. In other words, they can be viewed as the analogues of the frequencies in the angle-action representation. It follows that Eq. (5.34) has a structure similar to the resonance condition introduced in Chapter 3. As we saw in Section 3.8, resonances are "dangerous" in the sense that a small perturbation around the reference integrable system (here the free particle motion) is expected to give rise to hyperbolic points and unstable motions.

As long as l takes only discrete values, Eq. (5.34) will be satisfied only for certain p's. The collision operator will thus vanish almost everywhere in phase space. However, in the limit of a large system l becomes a continuous variable [Eqs. (5.32) to (5.33) become Fourier integrals]. As a result, it is expected that resonances will appear *almost everywhere* and that the collision operator will exist and be finite. We therefore are tempted to suggest that in systems in which the collision operator does not vanish, almost every phase space point is hyperbolic, as in the Baker transformation. Once again we see how irreversibility, which is inherent in kinetic theory, can be traced back to instability in the underlying dynamics.

5.10 RESONANCE AND LIGHT-MATTER INTERACTION

We have seen in Section 5.9 that resonances play a fundamental role in kinetic theory. It is only if they can be fulfilled almost everywhere in phase space that one can define the collision operator, the object at the basis of the modern theory of nonequilibrium processes. Resonances are thus responsible for the approach to equilibrium and for the irreversibility in the behavior of multibody systems.

In the present section we show that resonances play also a key role in the interaction between electromagnetic radiation and matter, one of the fundamental interactions of nature. However, in contrast to Section 5.9 where resonances appear automatically as soon as the system starts to evolve, here it will turn out that they are gradually created as a result of a self-organizing process of a new type.

Consider a harmonic oscillator of frequency ω. Its hamiltonian was given in Eq. (2.6b). We suppose that the oscillator is charged and embedded in an electromagnetic field, with which it interacts. As is usual in electromagnetic theory we describe the field as an infinite number of harmonic oscillators and assume that the frequencies of the latter are narrowly distributed around some value ω_e.

Owing to the presence of the field, the oscillator will experience a force proportional to its charge and depending on its velocity as well as on the intensity and orientation of the electric and magnetic fields. This force will induce an accelerated movement, even if initially the oscillator was at rest. Now, in classical electromagnetism it is shown that a charged particle undergoing an accelerated motion radiates, and thus loses, energy. The theory determining this "radiation damping" is not completely satisfactory, and for a more fundamental approach to the phenomenon of emission of radiation by matter one must resort to quantum mechanics. Nevertheless, the following equation of motion gives in a first approximation an idea of the role of resonance in light-matter interaction:

$$\frac{d^2x}{dt^2} + \frac{\gamma}{m}\frac{dx}{dt} + \omega^2 x = qF_e \cos \omega_e t \tag{5.35}$$

where x is the position coordinate of the oscillator, m and q are the mass and charge, respectively, γ is the damping constant, and F_e is the intensity of the field.

The solution of Eq. (5.35) can be obtained by standard methods in the form

$$x = \rho F_e \cos (\omega_e t + \theta) \tag{5.36}$$

with

$$\rho^2 = \frac{q^2}{(\omega^2 - \omega_e{}^2)^2 + \frac{\gamma^2}{m^2}\omega_e{}^2} \tag{5.37a}$$

$$\tan \theta = \frac{\gamma\omega_e}{m(\omega^2 - \omega_e{}^2)} \tag{5.37b}$$

We observe a very strong enhancement of the response when the intrinsic frequency ω of the oscillator approaches that of the field ω_e. In the limit of small damping γ the oscillator will therefore radiate essentially at the frequency of the field. This is precisely the phenomenon of resonance.

Let us now have a more fundamental look at the problem. We consider the full hamiltonian of the charged oscillator plus the field oscillators, including their interaction. We start with an initial condition corresponding to the charged oscillator in motion and the field oscillators at rest or, equivalently, in the state of zero energy (this is the classical vacuum, which when amended by quantum mechanics corresponds actually to a state of nonzero energy). We next solve on the computer the full (reversible) equations of motion, without making the phenomenological assumption of existence of a damping constant γ. The simulations then show that there is a short delay before the charged oscillator starts to radiate, which cannot be properly accounted by Eq. (5.35). It is very interesting to understand the origin of this delay. We have just seen that the radiation process involves a resonance between the material oscillator and the field oscillators. But at the start, there is no field oscillator available. As a first step the material oscillator has therefore to produce the field oscillators; or, in other words, to create the right environment in order to be able to radiate. Whatever the initial conditions, after a short time (of the order of 10^{-14} to 10^{-18} second for the hydrogen atom), the oscillator starts to radiate. In short, the mechanism of radiation is an irreversible "autocatalytic" process which appears at the interface of vacuum with matter.

Let us compare this behavior with irreversible processes in fluids. There also, whatever the preparation, it is impossible to prevent the fluid from reaching thermal equilibrium. However, in this latter case, the time scale is of the order of a relaxation time (typically 10^{-9} second), which is much longer than the characteristic time of the irreversible processes preparing the emission of radiation.

The existence of relaxation processes in the very short time scales associated with the interaction of radiation with matter illustrates once more the universal role of irreversibility in nature.

5.11 CONCLUDING REMARKS

We live in a pluralistic world. Simple dynamical phenomena such as the periodic motion of the earth around the sun appear to us to be time reversible and deterministic. But we also find irreversible processes such as chemical reactions, and stochastic processes such as the choice between bifurcating states.

How is this possible? The classical view was that irreversibility and randomness on the macroscopic scale are artifacts due to the complexity of the collective behavior of intrinsically simple objects. The belief in the simplicity of the fundamental level was one of the driving forces of classical science for nearly three centuries.

Today we must reappraise the situation. The fundamental level is not simple, not even in classical physics, which was the stronghold for determinism and reversibility for such a long time. The classical point of view emerged from the study of systems in which periodic motion and elliptic points played the essential role. In contrast, we now see that there exists a multitude of systems in which hyperbolic points play the central role. In this chapter, we have indicated two types of systems, K-flows and systems studied in modern kinetic theory, in which almost all points in phase space are hyperbolic. For such systems the concept of trajectory loses its observational support, and dynamics has to be embedded in a broader formalism that includes stochasticity and irreversibility. This step, which leads from groups to semigroups, is not due to our "ignorance"; it is due to the very structure of the dynamical systems.

In physics it is often useful to distinguish between states and laws. States can only be known through the preparation and observation of a system; and the specification of these processes is an indispensable part of any objective description of the system, as stressed in particular by Niels Bohr and Leon Rosenfeld. This preparation and observation can only lead to a "finite window." The mechanism of our measurement cannot be different from any other natural interaction between dynamical systems. It therefore includes the finite window associated with their limited sensitivity.

The recognition of this finite window leads to a new relation between states and laws. In the classical view (including that of quantum mechanics) states are time symmetric and are propagated by laws that preserve this time symmetry. Now we have to consider states such as the Boltzmann ensembles introduced in Section 5.8, which have broken time symmetry and are propagated by laws with broken time symmetry.

Complexity and the Transfer of Knowledge

Understanding what is going on around us is equivalent to building models and confronting them with observations. This statement may sound like a truism to a physicist or a chemist, but it goes far beyond physics and chemistry. At each moment our sensory systems scan the surroundings, the brain registers and compares the observations with respect to images already formed, and eventually reaches a preliminary conclusion. One of the basic steps in this procedure is the extensive use of *analogies* and *archetypes*.

Our purpose in this chapter is to show that physico-chemical systems giving rise to transition phenomena, long-range order, and symmetry breaking far from equilibrium can serve as an archetype for understanding other types of systems that show complex behavior—systems for which the evolution laws of the variables involved are not known to any comparable degree of detail. More important, in many of these systems the very choice of what should be a pertinent variable may well be part of the problem we try to solve.

In all the examples to be discussed, the analysis proceeds in two steps. First, certain analogies are drawn between the observations and the behavior of reference physico-chemical systems. This defines the type of model that is likely to be the most adequate representation of the system of interest. Next, we try to go beyond the stage of simple analogies, to recognize, *within the framework of the model adopted*, the specificities of each problem and to incorporate them in the description. The results of the analysis are finally confronted with experience, and if qualitative agreement is obtained, they are used to make predictions. We hope that concrete suggestions on how to *master complexity* will emerge as the natural outcome of this process.

6.1 NONLINEAR DYNAMICS IN FAR-FROM-EQUILIBRIUM CONDITIONS AND THE MODELING OF COMPLEXITY

Throughout this book, complexity has been connected to the ability to switch between different modes of behavior as the environmental conditions are varied. The resulting flexibility and adaptability in turn introduce the notion of *choice* among the various possibilities offered. It has been stressed that choice is mediated by the dynamics of fluctuations and that it requires the intervention of their two antagonistic manifestations: short-scale randomness, providing the innovative element necessary to explore the state space; and long-range order, enabling the system to sustain a collective regime encompassing macroscopic spatial regions and macroscopic time intervals.

A necessary prerequisite of all these phenomena is a nonlinear dynamics that gives rise, under suitable constraints, to *instability of motion* and to *bifurcations*. The first step in modeling complex behavior is therefore to assess the nonlinear character of the underlying dynamics and to identify a set of variables capable of showing instabilities and bifurcations. This procedure is greatly facilitated by the results of the theory of dynamical systems surveyed in Chapter 3 (see especially Section 3.6), according to which the essence of the qualitative behavior is often contained in simple universal evolution equations, the *normal forms*, satisfied by a limited set of collective variables, the *order parameters*. Potentially, therefore, a multivariable system, like most of the systems encountered in practice, may reduce to a simplified description near a bifurcation point. For instance, if for some reason we know that the system of interest may undergo transitions to two simultaneously stable steady states without space symmetry breaking, we are tempted to use a cubic dynamics, Eq. (3.14), as zero-order description. If, on the other hand, we are in the presence of a transition to a sustained oscillatory behavior, we would be entitled to regard the system of Eqs. (3.16) for the radial and phase variables as a promising starting point. Finally, a chaotic dynamics would suggest that a continuous description should involve at least three variables and that, in a discrete description afforded, for instance, by a Poincaré map, we might look for an iterative dynamics similar to the one represented by Figure 59 and Eqs. (3.31).

These considerations, whose usefulness should not be underestimated, must naturally be used with lucidity, otherwise they may overshadow what is—and must remain—an important aspect of the problem, namely the specificity of the system considered. Therefore, at each stage of the analysis the modeler must be able to relate the order parameters to the variables originally present, to attach the normal-form dynamics to the mechanisms of the processes going on in the system, and to satisfy the constraints that may be imposed by the very nature of the problem. In systems involving physico-chemical variables, these may be the positivity of the entropy production, of the temperature, or of the concentrations. In population problems the constraints will likewise impose the condition that the population of a species can never become negative. And in geophysical problems they will select models in which the space and the time scales revealed by the data are adequately reproduced by the form of the rate functions and the parameter values adopted.

Bearing these points in mind, we will now proceed to a series of examples in which modeling using nonlinear dynamical systems far from equilibrium as an archetype has been applied successfully. We consider, successively, materials science, biology, geophysics, and the social behavior of animal and human populations.

6.2 MATERIALS SCIENCE

One domain in which the impact of developments in nonlinear dynamics in far-from-equilibrium conditions is rapidly increasing is the science of materials. The synthesis of new materials is a crucial part of new technological developments. It is striking to realize that in many cases it is conditioned by processes operating at a great distance from thermodynamic equilibrium.

For instance, it is now realized that the transport phenomena going on in the host fluid phase during the growth of crystals and the solidification of alloys have a deep influence on the structure and the qualities of the solid phase. As in many problems involving fluids under nonequilibrium constraints, convection plays a dominant role because it affects the temperature and composition at the interface between fluid and solid phases. The convective movement of the fluid is in turn determined by the competition between external conditions (forced convection) and internal processes (natural convection). The latter are due to concentration or temperature differences (Section 1.3), but they can also be induced by surface tension (Section 1.6). Crystal growth and solidification are therefore coupled processes implying mass and heat transfer, fluid flows, chemical reactions, and phase transitions. As stressed in Section 1.3 and throughout Chapter 3, convection can become irregular under certain conditions. This chaotic behavior induces random variations of temperature and concentration, resulting in imperfections of the crystal structure.

The properties of the interface between liquid and solid are of course of primary importance in these phenomena. A striking manifestation of nonequilibrium in this context is the discovery that during crystal growth there exists a homogeneous phase between solid and liquid with properties that are different from those of the melt and the crystal. Light-scattering studies from this layer, whose dimensions are substantially larger than the lattice spacing, reveal a thermal diffusivity coefficient of the order of 10^{-8} cm^2 s^{-1}, as opposed to the typical value of 10^{-3} cm^2 s^{-1} characteristic of pure water. In other words, thermal disturbances are damped in a much less efficient way, a property suggesting that the entire layer is in a critical condition similar to that prevailing near a bifurcation point. Presumably, this transition is mediated by the heat flux through the interface arising from the advance of the solidification front, which sustains a literally *nonequilibrium phase of matter* (also called a mesophase) in the layer.

Another class of nonequilibrium phenomena arises as a result of irradiation when powerful laser or particle sources inject large amounts of energy in the surface layer of various materials during a very short interval of time. The fusion and resolidification phenomena that result are characterized by high values of the thermal gradients (10^6 to 10^8 K cm^{-1}) and of the interface velocities (1 m s^{-1}). These extreme nonequilibrium conditions are promising in the synthesis of new electronic materials, among other applications.

Perhaps the most striking intervention of nonequilibrium in the field of materials science pertains to the degradation of materials. Such phenomena as plasticity, deformation, or fatigue are usually mediated by defects that are generated by the stresses of nonequilibrium and that subsequently amplify and migrate to form precursor structures of fracture. Progress in this field requires an understanding of the formation of defects and dislocations, and of their interactions in systems subjected to external constraints. A modeling of the resulting nonlinear dynamics should lead to instabilities that induce multiple steady-state transitions and spatial structures for the defect concentration. Let us briefly outline some of the highlights.

We first identify the pertinent variables and parameters. In addition to the deformation, ε, which is the variable usually present in engineering approaches, the temperature, T, also seems to play a role. Indeed, during a tensile test a sample cools down while undergoing transformation in the elastic region. The amount of temperature drop is proportional to the deformation (assumed for simplicity to occur homogeneously within the bulk) and depends on the *Grüneisen parameter* γ. This latter quantity reflects the deviation of the intermolecular forces from the harmonic oscillator behavior and thus provides a measure of the thermo-mechanical coupling. At ordinary temperatures γ lies between 1.5 and 2.5 for a wide range of materials, but appears also to be highly strain-dependent, approaching a practically vanishing value for a critical magnitude of the applied constraint.

On the other hand, the decrease in temperature just noted is opposed by heating effects related to the viscosity of the material. As the applied stress increases, the initially static dislocations and other defects become more and more mobile, and their motion within the sample generates an increasing amount of thermal energy. This in turn further enhances the population of defects, not unlike what happens during an exothermic chemical reaction (Section 1.4). We are thus in the presence of a destabilizing mechanism with positive feedback, which counteracts the stabilizing effect of cooling induced by the deformation. The two mechanisms balance each other at some critical point, which can be used as a boundary between the elastic and plastic regimes of deformation. Figure 90 represents this remarkable coincidence. The threshold value σ_θ of the stress—the external constraint driving the system out of equilibrium—at this critical situation is expected to define a bifurcation point of a new state.

A more quantitative view of the above phenomena requires the derivation of a set of evolution equations for the key variables ε and T, in which the effect of the parameters σ and γ can be followed explicitly. We will not go into details, but only point out that under reasonable assumptions these two equations turn

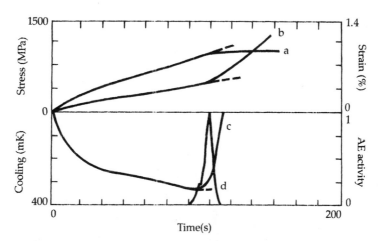

Figure 90 Variables and phenomena versus time for a steel sample. *a*, stress; *b*, deformation; *c*, temperature; *d*, acoustic emission activity. Notice the abrupt increase of temperature and acoustic activity in the critical region.

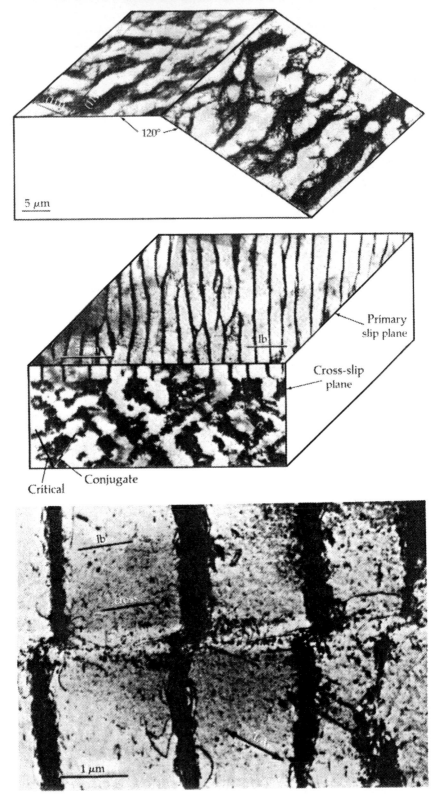

Figure 91 Three-dimensional dislocation structure in a periodically stressed single crystal of copper.

out to be coupled nonlinearly, through the term

$$-\tfrac{1}{2}C_v\left(\frac{\partial\gamma}{\partial\varepsilon}\right)\left(\frac{T-T_0}{T_0}\right)\varepsilon^2 \tag{6.1}$$

where C_v is the specific heat and T_0 is a reference temperature. Because of the strain dependence of the Grüneisen parameter γ noted above, this term is positive and introduces a mechanism of enhancement of the fluctuations leading to an instability. Near the bifurcation point σ_θ we can, following the results of Section 3.6, simplify the description further by eliminating all variables except the order parameter. The latter, which is a suitable linear combination of ε and $T-T_0$, obeys a dynamics of a form similar to Eq. (3.14) or the first expression in Eqs. (3.16):

$$\frac{dx}{dt} = (\sigma-\sigma_\theta)x - bx^3 \tag{6.2}$$

Depending on the conditions, the bifurcation can lead either to a new steady state, as in Eq. (3.14), or to a limit cycle oscillation, as in Eq. (3.16).

A more microscopic approach to the above problem requires us to analyze the balance between different elementary processes such as creation, annihilation, blocking, and diffusion of dislocations. This leads to reaction-diffusion models very similar to those used to study chemical instabilities (Section 3.11). When subjected to an anisotropic constraint, such models predict a hierarchy of symmetry-breaking transitions leading to dislocation patterns very similar to those observed experimentally, as shown, for example, in Figure 91.

In conclusion, it appears that mechanical phenomena of such current importance and concern as plasticity and yielding cannot be explored on purely mechanical grounds. Rather, they should be regarded as parts of the general problem of nonlinear dynamical systems functioning in far-from-equilibrium conditions. It is our belief that this realization itself constitutes a major breakthrough in the field of materials science.

6.3 THRESHOLD PHENOMENA IN CELLULAR DYNAMICS

Jumping from materials under stress to living cells might seem like a precarious enterprise, but such a process is at the heart of the art of modeling. We have just seen that plasticity and yielding can be viewed as threshold phenomena that are mediated by the competitive interactions between the defect population and the bulk material. When yielding does occur, we must conclude that the material has been unable to control the activity of the troublemaking defects, whose population has as a result reached catastrophic levels. It is the purpose of the present section to show that a very similar competition occurs in biological phenomena of vital importance, such as the growth of tumors. Here an initially small population of "troublemakers," say certain cells of a tissue, have for some of the reasons revealed by molecular biology lost their physiological function and become malignant. They subsequently tend to invade the organism by rapid proliferation, but the organism tries to counteract their action by sending specialized killer cells into the battlefield. The outcome of the resulting competition between malignant and killer cells will decide whether the troublemakers will be rejected or will become dominant. Analysis of this competition reveals what we believe to be a feature of the utmost importance, namely, that rejection or dominance can be viewed as a threshold phenomenon.

Before we describe the competition in more quantitative terms, we give a brief outline of the nature of the killer cells. The immune system, which provides the means by which vertebrates can counteract the action of pathogenic substances invading the organism, produces undifferentiated immune stem cells in the bone marrow. These cells differentiate subsequently into B- or T-lymphocytes and are released in the organism. Upon encountering an invading substance, usually referred to as an *antigen*, B-cells further differentiate into appropriate large-size cells that proliferate and secrete chemical substances capable of neutralizing the antigen, known as *antibodies*. On the other hand, T-cells, after further differentiation in the thymus, regulate the action of B-cells by both enhancement and suppression. In addition, they are involved in immune responses that are directly cell-mediated, particularly against cancer cells and foreign cells arising, for instance, from grafts. Other cellular species of the immune system, such as macrophages, have a similar function, referred to as cytotoxic activity. It is this cell-mediated response that will primarily concern us in the present section.

Let X be the population density of the proliferating target (e.g., malignant) cells, and E_0 the population density of the free cytotoxic cells. In accordance with experimental data, cells E_0 recognize the X's, fix them in the form of a complex $E \equiv E_0 X$, and subsequently lyse them through the dissociation of the E complex. This sequence of steps can be described by

$$
\left.
\begin{aligned}
X &\xrightarrow{\ \lambda\ } 2X \qquad \text{(proliferation)} \\
E_0 + X &\xrightarrow{\ k_1\ } E \xrightarrow{\ k_2\ } E_0 + P \qquad \text{(binding and lysis)}
\end{aligned}
\right\}
\tag{6.3}
$$

We assume that in the time scale of E_0 versus X competition, the state of the immune system can be regarded as stationary for all practical purposes. This means that the total population density of free and bounded immune cells, $E_t = E_0 + E$, remains constant in time. We furthermore adopt a simplified description whereby the kinetics of the two steps in the second of Eqs. (6.3) can be regarded as noncooperative (no feedbacks), at least in the range of parameter values of interest. We emphasize however that this assumption can easily be relaxed without affecting the results in a qualitative way.

Under these conditions, we may write the following balance equations:

$$dX/dt = \lambda X[1 - (X/N)] - k_1 E_0 X$$
$$dE_0/dt = -k_1 E_0 X + k_2 E$$

in the first of which the factor $(1 - X/N)$ expresses the existence of an upper limit N for X inside the volume element where the competition takes place. Since lysis is expected to be much faster than the other processes, the quasi steady-state approximation introduced in Section 3.6 can be invoked for the second equation. Setting $dE_0/dt \sim 0$ and using the conservation condition of the E species, we are able to express E_0 in terms of X and obtain

$$\frac{dX}{dt} = \left(1 - \frac{X}{N}\right)\lambda X - \frac{k_1 E_t X}{1 + \dfrac{k_1}{k_2} X} \tag{6.4}$$

This equation can be rewritten in dimensionless form as:

$$\frac{dx}{d\tau} = (1 - \theta x)x - \frac{\beta x}{1 + x} \tag{6.5}$$

with:

$$\beta = \frac{k_1 E_t}{\lambda} \qquad \theta = \frac{k_2}{k_1 N} \qquad \tau = \lambda t \qquad x = \frac{k_1}{k_2} X \tag{6.6}$$

It is obvious that Eq. (6.5) always admits the steady state solution $x = 0$. Linear stability analysis (see Appendix 1) shows that for $\beta < 1$ this solution is unstable while for $\beta > 1$ it is stable independently of the value of θ. The existence of a non-null steady state solution for x corresponds to tumoral states of the tissue. Two qualitatively different situations may now arise, according to the value of θ. They are represented in Figure 92. For $\theta > 1$, the existence of a tumoral state is restricted to values of $\beta < 1$. The condition for rejection is simply that in the course of the immune response the value of β must become larger than one. For $\theta < 1$, the transition between tumoral and nontumoral situations involves a bistability phenomenon whose properties are reminiscent of the nonequilibrium transitions discussed throughout this book. In the domain $1 < \beta < \beta_c = (1+\theta)^2/4\theta$, the null (normal) state and one tumoral state are simultaneously stable. The transition between these states is expected to involve a nucleation process. Starting with the tumoral state as an initial condition and increasing the value of β, rejection is ensured only for $\beta > \beta_c$.

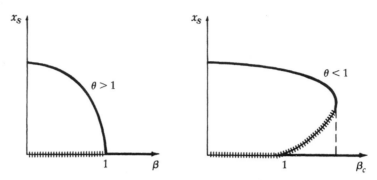

Figure 92 Steady-state concentration of tumoral cells plotted against the parameter β characterizing the activity of the immune system. In the righthand plot the possibility of the coexistence of a normal ($x_s = 0$) and a malignant ($x_s > 0$) state separated by a finite threshold is recognized.

An estimation of the parameter values based on data from the literature leads to the conclusion that in reality both types of steady states may arise. Remembering from the first relation in Eqs. (6.6) that the parameter β essentially describes the efficiency of the immune system, we realize that for a given state of the immune system in the range $1 < \beta < \beta_c$, $\theta < 1$, both normal and pathological situations may arise, depending on the system's history.

The dynamical view adopted in this section should provide an important element for the understanding of a host of other biological phenomena at the supermolecular level—for example, the aggregation of blood cells in arteriosclerosis.

6.4 MODELING CLIMATIC VARIABILITY

Populations of interacting entities will also be in the center of our preoccupations later in connection with the modeling of social behavior. Before going on to that problem, however, we present some ideas concerning the modeling of climatic variability, a problem of obvious concern not only at the scientific but also at the economic and societal levels. In spite of its global character, climatic variability remains basically a problem that should be amenable to a physico-chemical description, much like the problems discussed earlier in this chapter.

The major facts suggesting that the climatic system is characterized by a pronounced variability were surveyed in Section 1.8. Figure 93 sets the stage for the main purpose of the present section—the modeling of this phenomenon.

In the northern hemisphere at the peak of the last major glaciation, 18,000 years ago, a continental ice layer several kilometers thick covered much of northern America, down to the middle part of the United States, as well as much of northern and western Europe, to the latitude of Paris and Berlin. In the present-day

Figure 93 Schematic view of the northern hemisphere. Top, present-day conditions. Bottom, at the peak of the last glaciation.

situation, continental ice is confined essentially to Greenland. Furthermore, it is known that, as far as continental ice extent is concerned, the present situation was essentially established 10,000 years ago. In other words, we are forced to conclude that the planet earth, a physical system, has in a few thousand years (a short time on a geological scale) undergone a transition betwen two extraordinarily different states whose difference extends over the dimensions of the earth itself.

The above remarks suggest that any reasonable model of the climatic system should be able to account for the possibility of such large-scale transitions. Let us now survey the elements that we have at our disposal in order to build up an adequate model of this kind.

Climatic change essentially involves the solid earth surface, the oceans (or hydrosphere), the atmosphere, and the glaciers (or cryosphere). The state of each of these subsystems, as well as their interactions, are mediated by physical phenomena satisfying the basic balance equations of energy, mass, and momentum. Energy balance describes how the incoming solar radiation is processed by the system: what parts of that spectrum are absorbed and what are re-emitted, how they are distributed over different latitudes and longitudes as well as over continental and ocean surfaces. The most significant parts of mass balance are the balance of continental and sea ice and of some key minor atmospheric constituents like carbon dioxide and water vapor. Momentum balance determines atmospheric circulation, which eventually conditions local temperature and precipitation patterns. The problem is that all these balances are intimately coupled and the precise form of the phenomenological laws expressing the various fluxes present in terms of a limited set of macroscopic variables (see Section 2.2) is poorly known. Furthermore, one of the few undisputed realities is the existence of atmospheric and oceanic turbulence. If anything, this suggests (Section 3.10) that climate should share the intrinsic unpredictability of systems undergoing chaotic dynamics.

In the face of this situation, two attitudes can be adopted. On the one side, we can argue that the full set of balance equations in which individual fluxes are parameterized in the best possible way according to the data available, should be solved numerically on a computer. Work along this line gave rise to the *general circulation models* of meteorology and climatology and stimulated the development of ultra-fast computers capable of handling the enormous amount of input.

On the other hand, we can argue that precisely because of the enormous complexity of the climatic system, a *qualitative* analysis aiming to sort out long-term trends and to identify the key parameters is also needed. For the reader of this book, such an attitude should in fact appear to be the natural tool for handling complexity. After all, much of the progress in our understanding of phenomena at the laboratory scale—phenomena such as hydrodynamic and chemical instabilities, which are substantially simpler and easier to follow than global climate—is due to the extensive use of the concepts and methods of the theory of dynamical systems. In the remainder of this section, we present some examples of the kind of information and insight obtained by the qualitative analysis of simple mathematical models of climate. We focus on the problem of glaciation in the Quaternary era and, in particular, on the mechanisms by which the climatic system can capture and amplify the weak variations of incoming solar energy. As we saw in Section 1.8, despite their small values these variations seem to present an impressive correlation with the characteristic time scales of the glaciations.

Let us argue in very global terms, considering the earth as a zero-dimensional object in space receiving solar radiation and emitting infrared radiation back to space. In such a view there is only one important state variable, the mean temperature T, which evolves in time according to the heat balance equation

$$\frac{dT}{dt} = \frac{1}{C}\left\{Q[1 - a(T)] - \varepsilon_B \sigma T^4\right\} \tag{6.7}$$

Here C is the heat capacity of the system, Q the solar constant, and a the albedo, which expresses the part of the solar radiation emitted back to space. The temperature dependence of this quantity accounts for the surface-albedo feedback described at the end of Section 1.8.

The last term in Eq. (6.7) describes the emitted thermal radiation as a modification of the well-known Stefan-Boltzmann law of black body radiation. σ is the Stefan constant and ε_B is an emissivity factor accounting for the fact that the earth does not quite radiate as a black body.

A study of Eq. (6.7), summarized in Figure 94, shows that the system can admit up to three steady states. Two of them, T_b and T_a, are stable and correspond, respectively, to present-day climate and a cold climate reminiscent of a global glaciation. The third state, T_0, is unstable and separates the above two stable regimes.

As we saw in Section 4.5, the evolution of a system involving only one variable can always be represented as the evolution of a particle in a potential U which, in the present case, is defined by [see Eq. (4.30)]

$$U(T) = -\frac{1}{C}\int dT\left\{Q[1 - a(T)] - \varepsilon_B \sigma T^4\right\}$$

We call $U(T)$ the *kinetic potential*. Under the conditions of Figure 94, this potential is bistable, just as in Figure 75(b).

Now the climatic system, like any other physical system, is continuously subjected to statistical *fluctuations*, the random deviations from deterministic behavior discussed at length in Chapter 4. As we do not have sufficient information enabling us to write a master equation for the underlying probability distribution, we assimilate the effect of the fluctuations to a *random force*, $F(t)$, as discussed at

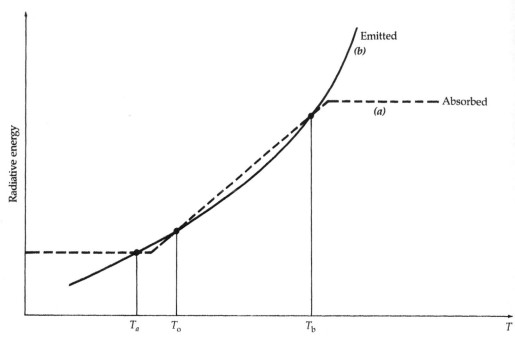

Figure 94 Global energy balance of the planet earth. With appropriate parameterization of the albedo function $a(T)$, curve (a) describes the temperature dependence of the absorbed part of incoming solar energy, and curve (b) describes the infrared radiation emitted by earth to space. These dependences allow for three steady-state regimes, two of which (T_a, T_b) are stable and correspond, respectively, to a glacial climate and a climate similar to that of the present day.

the end of Section 4.2. The energy balance equation, Eq. (6.7), now becomes a stochastic differential equation of the form

$$\frac{dT}{dt} = -\frac{\partial U(T)}{\partial T} + F(t) \tag{6.8}$$

The important new element introduced by this enlarged description is that different states become *connected* through the fluctuations. Stated differently, starting from some initial state the system will sooner or later reach any other state. This holds true as well for the stable states T_a and T_b, taken as initial states, whose deterministic stability is then transformed to some sort of *metastability*. As discussed in Section 4.5, the time scale of this phenomenon is determined by two factors: the *potential barrier*

$$\Delta U_{a,b} = U(T_0) - U(T_{a,b})$$

and the strength of the fluctuations as measured, typically, by the *variance*, q^2, of $F(t)$. The order of magnitude estimate given in Section 4.5, Eq. (4.31), yields:

Mean transition time from state T_a or T_b via the unstable state T_0

$$\tau_{a,b} \sim \exp\left(\frac{\Delta U_{a,b}}{q^2}\right) \qquad (6.9)$$

It is clear that if fluctuations are reasonably small, this time will be exceedingly large for any reasonable magnitude of the potential barrier. Typical values can be 10^4 to 10^5 years, that is to say, the range of the characteristic times of Quaternary glaciations. Still, we cannot claim that Eq. (6.9) provides an explanation of glaciation cycles, as the passage between the two climatic states T_a and T_b remains a random process, without a well-defined periodicity. It is here that the influence of a weak periodic external forcing becomes crucial.

Let us repeat the above analysis by adding to the solar constant Q a small term of the form $Q\varepsilon \sin \omega t$, whose period $\Delta = 2\pi/\omega$ represents the effect of variations affecting the mean annual amount of solar energy received by the earth, such as one of the orbital perturbations discussed in Section 1.8. As it turns out, if Δ is much smaller than the transition time $\tau_{a,b}$ the external forcing has practically no effect. But when Δ is comparable to $\tau_{a,b}$ the response of the system to the forcing is tremendously enhanced. Essentially, the presence of the forcing lowers the potential barrier and in this way facilitates the passage between climatic states. Moreover, because of the periodicity of the signal, the system is entrained and locked into a passage between two kinds of climate across the forcing-depressed barrier characterized by some average periodicity. We begin to see a qualitative mechanism of glaciations in which both internally generated processes and external factors play an important role. The simple model of *stochastic resonance* discussed above also allows us to identify the parameters that are likely to be the most influential in the system's behavior.

Naturally, the schematic character of the model calls for the development of more sophisticated descriptions which, while still amenable to an analytical study, incorporate additional key variables in the description. A particularly interesting class of such models comprises those describing the interaction between average surface temperature and continental or sea ice. For plausible values of the parameters they predict time-periodic solutions of the limit cycle type with periods comparable to the characteristic times of glaciation. However, by incorporating the effect of the fluctuations in the description, we find that any trace of coherence in the oscillating behavior will disappear after a sufficient lapse of time, owing to the weak stability properties of the phase of the oscillator. The coupling with a periodic external forcing allows for the stabilization of the phase and thus, once again, for a regime of entrainment and phase locking between the climatic system and the external forcing.

Further progress in the exciting field of climate dynamics should aim at diminishing the gap between simple qualitative models and detailed numerical descriptions. A re-examination of the climatic record in the light of the theory of dynamical systems is also expected to provide new insights, from which a novel description of complexity of our environment at large should emerge. Some attempts in this direction are summarized in Appendix 4.

6.5 PROBABILISTIC BEHAVIOR AND ADAPTIVE STRATEGIES IN SOCIAL INSECTS

We now turn to a different facet of complexity. We will be interested in phenomena in which the state variables utilized traditionally in the physical sciences, such as temperature or chemical concentrations, are no longer the variables by which complex behavior is manifested. New elements come into play and call for a language in which strategy, anticipation, symbols, and ritualization become the key words. Still, the possibility of performing transitions among various modes of behavior remains the principal fingerprint of complexity. This will allow us to apply in the analysis the concepts and methods of nonlinear dynamical systems functioning in far-from-equilibrium conditions. We begin our discussion with some aspects of the behavior of social insects.

Bees, ants, termites, and other social insects represent an enormous ecological success in biological evolution. They have conquered most of the known niches, particularly in the tropics, where their biomass can be as much as several hundred kilograms per acre.

The usual view of an insect society holds that nest construction, food searching, and other collective activities are the prototypes of a deterministic world in which individual insects are small, reliable automatons obeying a strictly established genetic program. Today, this picture of absolute rigidity is fading and a new paradigm is gradually emerging in which random elements from the environment and an adaptability of individual behavior begin to play an important role.

What is most striking in many insect societies is the existence of two scales: one at the level of the individual, characterized by a pronounced probabilistic behavior, and another at the level of the society as a whole, where, despite the inefficiency and unpredictability of the individuals, coherent patterns characteristic of the species develop at the scale of an entire colony. Let us see how these two aspects are related to ensure the overall organization of the society. We choose as a main example food searching by ants, and compare the course of this activity in two different situations. In the first, a unique, predictable food source such as a colony of tinier insect species (say aphids) whose lifetime is of the order of several months, exists in the vicinity of the nest. In the second situation, an unexpected source, such as a dead bird or a second colony of prey species, suddenly becomes available.

In the first case we observe the formation of stable trails from the nest to the nearby colony, each trail having its own specialized users; see Figure 95. Moreover, few ants are found outside these main axes. Clearly, under the existing

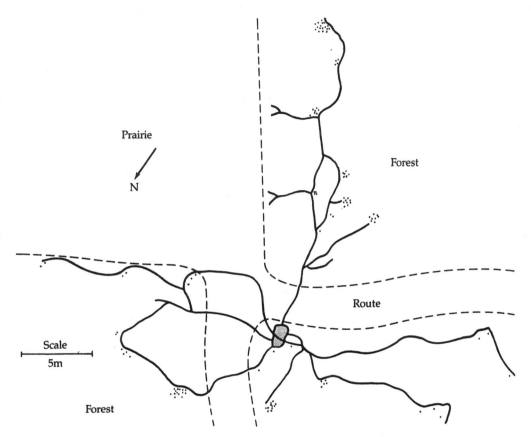

Figure 95 Schematic representation of the nest and the network of trails of a society of *Lasius fuliginosus.* ⌒ ant trail; ⬤ nest; ⠿ trees with aphids.

circumstances it is beneficial for the society to develop permanent stable structures with a low noise level, such as these trails and their patterns of use.

In contrast, a permanent structure in an unpredictable environment may well compromise the adaptability of the colony and bring it to a suboptimal regime. A possible reaction toward such an environment is thus to maintain a high rate of exploration and the ability to rapidly develop temporary structures suitable for taking advantage of any favorable occasion that might arise. In other words, it would seem that randomness presents an adaptive value in the organization of the society. This statement appears to be supported by the experimental data, to which we now turn our attention.

The discovery of a new food source requires the mobilization of ants to ensure its efficient exploration. The mechanisms by which ant societies manage to assemble great numbers of individuals around food sources constitute food *recruitment.* The food recruitment in two ant species, *Tetramorium impurum* and *Tapinoma*

Table 4 Comparison between recruitment accuracy in two ant species.*

	Tetramorium impurum		Tapinoma erraticum
Length (in %) of single recruitment trails actually followed by recruit	17(40)		67.7(47)
Percentage of recruits reaching food source	Alone	8.9(45)	73.6(216)
	In groups	60(10)	—
	Total	18.2(55)	73.6(216)

* Numbers in parenthesis are those of ants actually observed.

erraticum, has been studied. The first species can perform group recruitment: in returning from the food source to the nest the recruiter lays a trail and subsequently follows this trail to guide a small number of recruits from the nest to the source. Both species can perform mass recruitment, in which the trail, once laid, can guide recruits who travel independently of the recruiter.

Now both species are exposed to the same experimental conditions. Specifically, food recruitment is initiated by offering 1 mole sucrose solution at 10 cm from the nest after four days of starvation. The behavior of recruiters and recruits is videorecorded. The accuracy of recruitment is measured by the proportion of ants recruited who actually reach the food source, and also by the mean portion of a given recruitment trail actually followed by ants traveling alone. Table 4 summarizes the results. Clearly, a recruitment trail is a far more efficient stimulus for *Tapinoma* than for *Tetramorium*. Moreover, mass recruitment is much more accurate in *Tapinoma*. With the same experimental conditions, fully 73 percent of the recruits leaving the nest reach the source by following the trail laid by a single recruiter. In other words, the amount of error during food recruitment varies considerably between species. We now show that it can also vary within the same species, depending on the conditions.

Two containers ($5 \times 20 \times 30$ cm^3) are connected by a bridge. The nest is placed in one container, two food sources in the other. Recruitment toward sucrose solutions is initiated after eight days of starvation. Food sources are 11 cm distant from the bridge and 14 cm from each other. They are deposited simultaneously or with a one-hour interval between, and the traffic of *Tetramorium* ants on the bridge is videorecorded. As it turns out, recruitment is only slightly accurate. Only 20 to 40 percent of the recruits participate in the collection of food. Most of the ants lose the recruitment trail and explore the foraging area. They quickly discover the second source, even when recruitment toward the first source was already well established. Typically, when the two sources are of equal quality, one is exploited maximally until it is exhausted. Thereafter the second source is fully colonized, and its exploitation is intensified so that no interruption in food-collection rate is recorded. When the second food source is more concentrated in glucose than the first, the ants shift their collective efforts toward the most rewarding source, without however completely abandoning the first one. This behavior is strongly reminiscent of a bifurcation of two simultaneously stable states (see Figure 41) and of its deformation by the effect of a small imperfection (see Figure 42). Moreover, we see clearly how randomness allows the system to switch between these two modes of behavior.

We now ask the following question. Ants losing the trail because of errors are able to discover new food sources, but they do not exploit the already known ones. What is the best balance between fluctuations, which allow discoveries and innovations, and accurate determinism, which allows immediate exploitation? The following simple model provides some clues.

We first describe how global recruitment works in the presence of a single food source. Let X be the number of workers at the food source. The mean flux of ants arriving at the same source, J_+, may reasonably be set to be proportional to the number of encounters between the X working individuals and the $N - X$ remaining ones, where N is the total number of ants able to participate in the recruitment. Thus:

$$J_+ = aX(N - X) \tag{6.10a}$$

where a is the recruitment rate per individual. The flow of departure from the food source, J_-, is, on the other hand, merely

$$J_- = -bX \tag{6.10b}$$

where b is the inverse of the mean time spent to stay near the food and come back to the nest. When the food source is exhausted, this last term alone governs the evolution of X. But when the source still contains food the overall rate of change of X is given by:

$$\frac{dX}{dt} = aX(N - X) - bX \qquad \text{(food not exhausted)} \tag{6.11}$$

This relation is known as the *logistic equation*, and it has a widespread use in a large variety of population problems. The limiting value N is referred to as the carrying capacity. If $X < N$, J_+ is positive, and this expresses the positive feedback of X population into itself, albeit with a slowing down expressed by the saturation factor $N - X$.

Suppose now that the colony is confronted with k identical, regularly distributed food sources equidistant from the nest. Recruitment is initiated toward the median source, and it is assumed that ants are distributed according to a normal distribution around the source toward which they are recruited. Ants hitting a source are assumed to initiate recruitment toward it.

A straightforward generalization of Eqs. (6.10)–(6.11) yields:

$$\frac{dX_i}{dt} = \sum_{j=1}^{k} a_{ji} X_j (N - X_1 - \cdots - X_k) - bX_i \qquad (6.12a)$$

$$\text{(food source } i \text{ not exhausted)} \qquad i = 1, \ldots, k$$

$$\frac{dX_i}{dt} = -bX_i \qquad \text{(food source } i \text{ exhausted)} \qquad (6.12b)$$

where $X_1, \ldots, X_k$ now represent the number of recruited ants moving toward the source $1, \ldots, k$. The coefficient a_{ji}, $j \neq i$, is the recruitment rate for the ants recruited to j to reach by error source i. It can be estimated as:

$$a_{ji} = \frac{1}{S} \int_{|r_i - r| \leqslant d} dr \exp\left(-\frac{(r_j - r)^2}{2\sigma^2} \right) \qquad (6.13)$$

in which S is a suitable normalization factor, $2d$ is the diameter of the food source, σ^2 is the variance of the gaussian probability distribution, and r_i is the location of source i. The smaller the σ the more accurate the recruitment.

Figure 96 describes the results of a numerical solution in which the time needed to fully exploit a given quantity of the food is shown as a function of the error as expressed by the variance of the statistical distribution, for given numbers of food sources. As expected, when only one source is present the best recruitment strategy is the one that functions without mistake. However, when food is divided among several locations, there is one optimal value of the error which minimizes the time of exploration. This defines the level of randomness, which can be advantageous by increasing the possibility of discoveries, thus allowing the society to concentrate its activities on the most rewarding resources and to promote the colonization of resources that will be fully exploited later.

So far in our discussion individual ants were differentiated only by the fact that some aspects of their behavior were distributed statistically. However, there is a division of work inside an insect society. An ant nest is a female society, in which the role of the short-lived males is limited to fecundation. Many popular books describe the division of work between queens, which ensure reproduction, and the workers, which carry out the tasks necessary for the functioning of the society.

In certain species the group of workers is composed of many physically distinct castes, and a division of work based on age is well established. Surprisingly, however, it seems that in a given caste, within a given age group, workers manifest marked differences of activity whose significance for the society is not clear.

When ants find themselves in unfavorable conditions, they move their nests. The moving proceeds by exploration, discovery of a favorable site, and transport of the content of the previous nest. An experiment has been designed in which a colony has been subjected to a series of movings in a period of eight months,

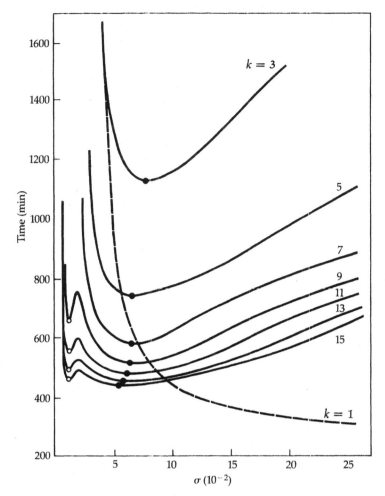

Figure 96 Collection times for different number of food sources, k. Parameter values: $N = 650$; $a = 10^{-3}$; $b = 16 \times 10^{-3}$; $2d = 1.4$ cm.

and the number of charges transported by each of the members has been measured. A most surprising result is that there seems to be a certain stability in the level of activity: ants recorded to be active in a first experience will tend to remain active for several months, and similarly for "lazy" ants: they remain less active. These differences are not entirely intrinsic to the individuals, but appear to be modulated by the environment and by the history of the particular individual. Indeed, when the two groups are separated, lazy ants become active. If the two groups are again merged, the colony will keep the memory of the "perturbation" of separation for a certain time.

In conclusion, an insect society appears to possess a remarkable adaptability that allows it to perform, with minimum programming at the genetic level, a series of complex tasks. Once more, transitions to different modes of behavior triggered by the environmental conditions appear to be the primary mechanism at the basis of this extraordinary performance.

6.6 SELF-ORGANIZATION IN HUMAN SYSTEMS

Our everyday experience teaches us that adaptability and plasticity of behavior, two basic features of nonlinear dynamical systems capable of performing transitions in far-from-equilibrium conditions, rank among the most conspicuous characteristics of human societies. It is therefore natural to expect that dynamical models allowing for evolution and change should be the most adequate ones for social systems.

A dynamical model of a human society begins with the realization that in addition to its internal structure, the system is firmly embedded in an environment with which it exchanges matter, energy, and information. Think, for instance, of a town in which raw materials and agricultural products arrive continuously, finished goods are exported, while mass media and professional communication keep the various local groups aware of the present situation and of the immediate trends.

The evolution of such a system is an interplay between the behavior of its actors and impinging constraints from the environment. It is here that the human system finds its unique specificity. Contrary to the molecules, the actors in a physico-chemical system, or even the ants or the members of other animal societies, human beings develop individual *projects and desires*. Some of these stem from anticipations about how the future might reasonably look and from guesses concerning the desires of the other actors. The difference between desired and actual behavior therefore acts as a constraint of a new type which, together with the environment, shapes the dynamics. A basic question that can be raised is whether, under those circumstances, the overall evolution is capable of leading to some kind of global optimum or, on the contrary, whether each human system constitutes a unique realization of a complex stochastic process whose rules can in no way be designed in advance. In other words, is past experience sufficient for predicting the future, or is a high degree of unpredictability of the future the essence of human adventure, be it at the individual level of learning or at the collective level of history making? The developments outlined in the preceding chapters suggest that the answer to this question should lean toward the second alternative. Let us see whether the mathematical modeling permits us to justify this intuitive feeling and, at the same time, to specify more sharply the nature of the unpredictability of the human system.

We first attempt to estimate the payoff associated with a given choice. We may suppose that the desirability of making a particular decision i from among K alternatives, per unit time, is proportional to the relative attractiveness of the

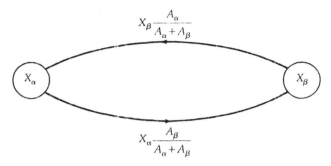

Figure 97 Feedback loop describing choice under the influence of two simultaneously available options, α and β, whose attractiveness is A_α and A_β, respectively. X_α, X_β: populations choosing α or β at a given time.

decision. However, as a given option is gradually adopted, the payoff will in general change, and so the pattern of preferences at the level of the population will reflect this as choices become more or less attractive. This feedback is shown in Figure 97 in the particular case of two choices, α and β. Here A_α represents the attractiveness of choice α and X_α is the number of people having adopted α at a given time. Clearly, the relative number of individuals wishing to switch to choice β will be proportional to the number of those having adopted some other choice, like α, multiplied by the relative attractiveness of β: $A_\beta/(A_\alpha + A_\beta)$. Similarly, those individuals wishing to leave choice β in favor of α will be proportional to X_β multiplied by the relative attractiveness of α: $A_\alpha/(A_\alpha + A_\beta)$. This results in a set of balance equations for X_α and X_β that look very much like the ecological equations (6.11)–(6.12) obtained in the previous section:

$$\frac{dX_\alpha}{dt} = aX_\alpha \left(\frac{X_\beta A_\alpha}{A_\alpha + A_\beta} - \frac{X_\alpha A_\beta}{A_\alpha + A_\beta} \right)$$

or, realizing that $X_\beta = N - X_\alpha$, N being the total population size:

$$\frac{dX_\alpha}{dt} = aX_\alpha \left(\frac{NA_\alpha}{A_\alpha + A_\beta} - X_\alpha \right) \tag{6.14}$$

and a similar equation for X_β. Comparing with Eq. (6.11), we see that the existence of options affects the carrying capacity of the system and makes it, through the dependence on the attractiveness on the X's, a function of the instantaneous state.

The above discussion can easily be extended to an arbitrary number of choices K, and to the more realistic situation in which the attractiveness of an option i depends on the particular population j envisaging this option. Thus we arrive at the following set of equations:

$$\frac{dX_i}{dt} = CX_i \left(1 - \frac{X_i}{\sum\limits_{j=1}^{K} \dfrac{N_j A_{ij}}{\sum\limits_{l} A_{lj}}} \right) \qquad (i = 1, \ldots, K) \qquad (6.15)$$

where it is further assumed that the total population is inhomogeneous and thus breaks to distinguishable subpopulations N_j, each having its own viewpoint about the relative attractiveness of the choices. The N_j's satisfy a set of equations similar to the equations of population dynamics, see Eqs. (6.12).

The behavior of the dynamical system described by Eqs. (6.15) depends crucially on the particular way in which the attractiveness A_{ij}'s and the N_j's will depend on the population variables X_i, which characterize the instantaneous state. This in turn depends on the particular system under consideration. In the following we discuss this problem in connection with the evolution of urban structures.

Consider the development of an urban center as certain areas specialize in specific economic activities and as residential segregation produces neighborhoods differing in their living conditions and access to jobs and services. Two major components of the population are recognized, corresponding essentially to blue- and white-collar workers. As for the jobs, they may belong to industry (export or local), so-called tertiary functions (basic and specialized services), or financial activities. The principal variables determining the state are now the number of residents X_i^k of group k ($k = 1, 2$) at a point i, and the corresponding jobs J_i^k. The different "options" offered to the individuals are located at points i, which are thought of as sites of a lattice representing the space available. The interaction mechanisms between the different variables are summarized in Figure 98.

To write the modified form of Eqs. (6.15) for X_i^k, we must express the quantities N_i and A_{ij} in terms of J_i^k and X_j^k. It may be reasonably assumed that the subpopulation N_i is essentially determined by the number of jobs of type k situated at point i. As for the attractiveness A_{ij}'s, they are complicated functions of X_i^k, J_i^k and of the distance separating the points i and j. Their explicit form need not be specified here, but we point out that it includes such considerations as cost and time to travel to work, price of land, character of the neighborhood, and similar factors.

Similar equations can be written for the other categories of actors, equations that describe, for example, the need for industrial employment to be located at a point with good access to the outside and, preferably, in a location adjacent to already established industry.

The model just defined views the evolution as an autonomous process whose course is determined at each moment by the mechanisms of interaction among different actors. The environmental constraints act through the parameters, and the initial condition can be adopted to express the effect of randomness or of a sys-

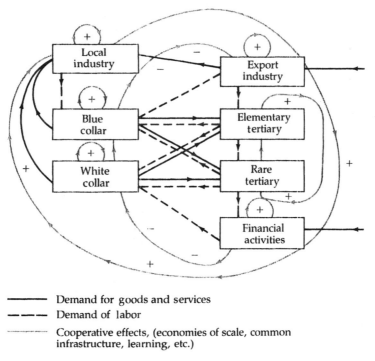

Demand for goods and services
‑‑‑‑‑‑ Demand of labor
Cooperative effects, (economies of scale, common
infrastructure, learning, etc.)

Figure 98 Schematic representation of the interactions and the flows of information in a simplified model of a city.

tematic external intervention or "planning." An alternative scenario, closer to what happens in reality, is to let the system evolve for a certain period of time, brutally modify its state by launching a new activity or an "innovation," again let the system follow its autonomous dynamics until a new innovation is launched, and so forth. As the equations are highly nonlinear, it is expected that there will be several solution branches exhibiting a complicated set of bifurcation phenomena. Different initial conditions will place the system in different basins of attraction, thus switching on different evolutions, different histories. Recording a particular history among the multitude of the possible histories does not necessarily reflect the action of a global planner attempting to optimize some overall function, but simply that this particular pattern is a stable and viable mode of behavior. We now have sufficient elements at our disposal to answer the question raised at the beginning of the present section.

Numerical solution of the equations of the model establish unequivocally the existence of a large number of solutions and of intricate bifurcation phenomena. Starting from a space in which variables are initially distributed at random, we observe the gradual emergence of an organized pattern with its own administrative and business centers, its industrial zones, its shopping centers, and its residential neighborhoods of varying qualities. In the absence of any massive disturbance the pattern remains stable indefinitely. This spontaneous symmetry breaking is very similar to the formation of spatial structures in hydrodynamics and chemical kinetics reviewed in Sections 1.3 and 1.4.

A very interesting result emerging from the model is the following. If a new activity is launched at a certain time, it will grow and stabilize. If the place is well chosen, it may even prevent the success of similar attempts made nearby at a later time. However, if the same activity is launched at a different time, it need not succeed; it may regress to zero and represent a total loss. This illustrates the dangers of short-term, narrow planning based on the direct extrapolation of past experience. Such static methods threaten society with fossilization or, in the long term, with collapse. The principal message of the dynamical modeling advocated in this section is that the adaptive possibility of societies is the main source allowing them to survive in the long term, to innovate of themselves, and to produce originality.

Linear Stability Analysis

A1.1 BASIC EQUATIONS

In Sections 2.6 and 2.7 the concepts of stability and bifurcation were defined, while in Chapter 3 a number of general properties of the branching of solutions and their stability were discussed. This appendix is addressed to the mathematically more sophisticated reader; here the formal method for computing the stability of the solutions of a dynamical system is outlined. In Appendix 2 this formalism is completed by an outline of the method for computing solutions that bifurcate beyond the instability.

We have repeatedly stressed that the property of stability refers to the response of the system to perturbations of various kinds. We imagine therefore a certain reference state, $X_{1s}, \ldots, X_{is}, \ldots$, where $\{X_i\}$ represents the set of state variables, which is continuously probed by internal fluctuations or external disturbances. If we deal with a completely unbiased system, free of external forces and subjected to a completely uniform and time-independent environment, $\{X_{is}\}$ will naturally represent the time-independent and spatially uniform solution that is expected to exist under such conditions, and which describes the absence of self-organization and complex behavior within the system. More generally, we may choose $\{X_{is}\}$ to be the solution having the highest symmetries allowed by the external constraints. In the terminology of Section 2.7 $\{X_{is}\}$ lies on the *thermodynamic branch*, the continuation of near-equilibrium behavior as the distance from equilibrium is increased. Self-organization and the onset of complex behavior can thus be viewed as a transition from $\{X_{is}\}$ to solutions of a new type.

Let us now set up an appropriate formalism to test the stability of the reference state. Much of the difficulty of stability theory lies in the presence of a great number of variables characterizing a problem, and in the fact that in many cases these variables may be functions of the spatial coordinates. We therefore introduce a shorthand notation in which the entire set of variables is represented by a column vector X, whose components are $X_1, \ldots, X_i, \ldots$. In a space-independent situation, the total number n of variables (e.g., chemical concentrations) is typically finite. X will then be defined in the usual vector space familiar from algebra. But if some of the variables depend on the space coordinate r in addition to the above vector, X will also lie in a functional space, for example, the space of functions whose squares are integrable, which is familiar from elementary quantum mechanics.

Using this notation we can write the rate of change of X in the following form [cf. Eq. (2.12) or (3.1)]

$$\frac{\partial X}{\partial t} = F(X, \lambda) \tag{A1.1}$$

Here F is an *operator* acting on the space in which X is defined. It is generally nonlinear, as a result of the feedbacks acting on the system. λ denotes a set of control parameters affecting the evolution, such as rate constants, diffusion coefficients, and so on.

The reference state X_s introduced above is itself a particular solution of Eq. (A1.1), hence

$$\frac{\partial X_s}{\partial t} = F(X_s, \lambda) \tag{A1.2}$$

The role of perturbations on stability will be expressed by setting

$$X = X_s + x \tag{A1.3}$$

where x represents the disturbance, and by converting Eq. (A1.1) into an equation for x:

$$\frac{\partial x}{\partial t} = F(X_s + x, \lambda) - F(X_s, \lambda) \tag{A1.4}$$

Some examples will help the reader realize the way x and the operator F are to be understood. The simplest case is certainly a problem involving a single, space-independent variable, such as the model of Eq. (2.15) in a well-stirred medium. F reduces then to a scalar:

$$F = \text{Reaction rate} = -k_2X^3 + k_1aX^2 - k_3X + k_4b$$

while the righthand side of Eq. (A1.4) becomes:

$$\begin{aligned}
F(X_s + x, \lambda) &- F(X_s, \lambda) \\
&= -k_2(X_s + x)^3 + k_1a(X_s + x)^2 - k_3(X_s + x) + k_4b \\
&\quad - (-k_2X_s^3 + k_1aX_s^2 - k_3X_s + k_4b)
\end{aligned}$$

or, after expanding the cubic and quadratic binomials:

$$\frac{\partial x}{\partial t} = (-3k_2X_s^2 + 2k_1aX_s - k_3)x + (-3k_2X_s + k_1a)x^2 - k_2x^3 \tag{A1.5}$$

If, on the other hand, we take the brusselator model in a well-stirred medium, Eq. (3.18), we obtain the following two-component vector representation of the state variables and rate functions:

$$X = \begin{pmatrix} X \\ Y \end{pmatrix}, \quad X_s = \begin{pmatrix} A \\ B/A \end{pmatrix}, \quad F = \begin{pmatrix} A - BX + X^2Y - X \\ BX - X^2Y \end{pmatrix}$$

Straightforward algebra leads then to the following explicit form of Eqs. (A1.4):

$$\frac{\partial}{\partial t}\begin{pmatrix} x \\ y \end{pmatrix} = \begin{bmatrix} (B - 1)x + A^2y + \dfrac{B}{A}x^2 + 2Axy + x^2y \\[2mm] -Bx - A^2y - \dfrac{B}{A}x^2 - 2Axy - x^2y \end{bmatrix} \tag{A1.6a}$$

If the reactions are taking place in an unstirred medium, F will comprise an additional contribution arising from the transport of matter through diffusion or convection. For the particular class of reaction-diffusion systems, whose properties were surveyed at length in Section 1.4, only diffusion needs to be taken into account. As stressed in Section 2.2, Fick's law, Eq. (2.7b), frequently provides an excellent representation of this process. The corresponding operator F, for instance for the brusselator, will then be:

$$F_{dif}(X) = \begin{pmatrix} D_1 \nabla^2 X \\ D_2 \nabla^2 Y \end{pmatrix}$$

Because of the linear dependence on x and y this leads to

$$\frac{\partial}{\partial t} \begin{pmatrix} x \\ y \end{pmatrix}_{dif} = \begin{pmatrix} D_1 \nabla^2 x \\ D_2 \nabla^2 y \end{pmatrix} \tag{A1.6b}$$

Note that F_{dif} is acting both on the space of two-dimensional vectors and on the space of functions depending on the coordinate r.

Let us return to the general formalism, Eqs. (A1.4). As suggested by the explicit examples treated above, it is natural to expand the righthand side around the reference state X_s. If F has a polynomial structure in X this is always possible, and leads to a finite number of terms. In more intricate situations, however, F may depend differently on X. To tackle the problem in this case we assume that:

1 F can still be expanded in powers of x
2 The expansion may be truncated to finite order

This latter assumption limits us to the study of *infinitesimal stability*, that is to say, the response of the system to small disturbances such as $|x|/|X_s| \ll 1$. This is still very useful, since infinitesimal stability gives necessary conditions for instability, in the sense that if X_s is unstable toward small x it will be unstable toward any x.

Formally, the above expansion can be represented as:

$$F(X_s + x, \lambda) = F(X_s) + \left(\frac{\delta F}{\delta X}\right)_{X_s} \cdot x$$
$$+ \tfrac{1}{2} \left(\frac{\delta^2 F}{\delta X \, \delta X}\right)_{X_s} \cdot xx + \tfrac{1}{6} \left(\frac{\delta^3 F}{\delta X \, \delta X \, \delta X}\right)_{X_s} \cdot xxx + \cdots \tag{A1.7}$$

The objects $(\delta F/\delta X)_{X_s}$, etc. are generalizations of derivatives, and are referred to in the mathematical literature as Fréchet derivatives. As an example, a comparison with Eqs. (A1.6a)–(A1.6b) leads to the following identification of the first Fréchet derivative for the brusselator model:

$$\left(\frac{\delta F}{\delta X}\right)_{X_s} = \begin{pmatrix} B - 1 + D_1 \nabla^2 & A^2 \\ -B & -A^2 + D_2 \nabla^2 \end{pmatrix} \tag{A1.8}$$

showing that $(\delta F/\delta X)_{X_s}$ acts as a matrix in the space of two-dimensional vectors and as a differential operator in the space of space-dependent functions.

A1.2 THE PRINCIPLE OF LINEARIZED STABILITY

Equations (A1.4) together with Eq. (A1.7) constitute a highly nonlinear problem which, as a rule, is as intractable as the original problem, Eq. (A1.1). At this point, however, a most important result of analysis can be invoked to enable further progress. This theorem, also known as the principle of linearized stability, compares the stability properties of the following two problems:

- The original, fully nonlinear problem
- The linearized problem, in which higher order terms are omitted

Introducing the shorthand notations

$$\mathscr{L}(\lambda) = (\delta F/\delta X)_{X_s}$$
$$h(x, \lambda) = \tfrac{1}{2}\left(\frac{\delta^2 F}{\delta X \delta X}\right)_{X_s} \cdot xx + \cdots \tag{A1.9}$$

for the linear and nonlinear contributions, respectively, we can formulate these two problems as the solutions of the following sets of equations:

$$\frac{\partial x}{\partial t} = \mathscr{L}(\lambda) \cdot x + h(x, \lambda) \qquad \text{(nonlinear problem)} \tag{A1.10}$$

and

$$\frac{\partial x}{\partial t} = \mathscr{L}(\lambda) \cdot x \qquad \text{(auxiliary linearized problem)} \tag{A1.11}$$

Notice that both systems are homogeneous, being satisfied identically by the trivial function $x = 0$.

The theorem of linearized stability now stipulates the following:

1. If the trivial solution $x = 0$ of the linearized problem, Eq. (A1.11), is asymptotically stable, then $x = 0$ (or equivalently $X = X_s$) is an asymptotically stable solution of the nonlinear problem, Eq. (A1.1) or (A1.10).

2. If the trivial solution $x = 0$ of the linearized problem, Eq. (A1.11), is unstable, then $x = 0$ (or equivalently $X = X_s$) is an unstable solution of the nonlinear problem.

We do not reproduce the proof here, but in view of the comments made in the last part of Section A1.1 we should now have an intuitive feeling about the reasonableness of the assertions of the theorem. The important point is that, by virtue of the theorem, stability, one of the fundamental properties of dynamical systems, is reduced to a linear problem—a much more traditional and tractable problem of analysis.

A1.3 THE CHARACTERISTIC EQUATION

To proceed further it will be instructive to focus on the particular class of *autonomous systems* for which the constraints are time-independent, and to suppose that the reference state X_s is a steady-state solution. It follows that the linear stability operator $\mathscr{L}(\lambda)$ is time-independent. Under these conditions, it can easily be checked that Eqs. (A1.11) admit solutions of the form:

$$x = ue^{\omega t} \tag{A1.12}$$

In other words, the time dependence of the solutions is contained entirely in the exponential function (note that ω is generally complex-valued), whereas the vector u takes into account the structure of x as a vector in the space of the state variables as well as its dependence on the spatial coordinate r.

Substituting into Eqs. (A1.11) and remembering that $\mathscr{L}(\lambda)$ is time-independent, we realize that the exponential $e^{\omega t}$ cancels on both sides and we are left with

$$\mathscr{L}(\lambda) \cdot u = \omega u \tag{A1.13}$$

This relation, supplemented with appropriate boundary conditions, defines an *eigenvalue problem*, a classical problem of analysis. Such problems arise frequently in quantum mechanics or in heat and mass transfer problems. What is remarkable is that, independently of the properties of the eigenvectors u, once the eigenvalues ω are determined the stability problem is solved. Indeed, remember from Eq. (A1.12) that x varies in time as

$$x \sim e^{\omega t} = e^{(Re\omega)t} \cdot e^{i(Im\omega)t}$$

where $Re\omega$ and $Im\omega$ are, respectively, the real and the imaginary parts of the complex-valued quantity ω. It follows therefore that if $Re\omega < 0$, x is an exponentially decreasing function (with or without an oscillatory modulation according to whether $Im\omega$ is nonvanishing or vanishing). Hence the reference solution $x = 0$ is reached in the limit $t \to \infty$, in other words, $x = 0$ is asymptotically stable. If, on the other hand, $Re\omega > 0$, the perturbations are growing exponentially, and the reference solution $x = 0$ is unstable.

These two regimes, for which the principle of linearized stability applies, are separated by a regime known as *marginal stability*, $Re\omega = 0$, which signals the threshold of instability of the reference state.

We can now understand better the importance of the concept of the control parameter λ. Indeed, a variation of λ induces a variation of $\mathscr{L}$ and, through it, of the eigenvalue ω. The existence of a transition between two qualitatively different regimes will then be reflected by the fact that at least one of the eigenvalues ω will change as a function of λ in the form depicted in Figure A.1 (also see Figure 48). The value λ_c of λ at which $Re\omega$ will change sign is the *critical value* beyond which an instability is bound to occur.

The specific dependence of the eigenvalue ω on the control parameter λ is determined by the problem considered. A very important class of systems, which we have encountered frequently, comprises systems in which the operator F can be decomposed into two parts, one expressing the rate $v(x, \lambda)$ of a local process like a chemical reaction, and another expressing the transport of matter through diffusion. A particular illustration of these *reaction-diffusion* systems is the brusselator, Eqs. (A1.6a)–(A1.6b).

More generally, for a multivariable reaction-diffusion system, we write

$$F = v(X, \lambda) + D \cdot \nabla^2 X \qquad (A1.14)$$

whose linearized operator $\mathscr{L}$ is given by

$$\mathscr{L} = (\partial v / \partial X)_{X_s} + D \cdot \nabla^2 \qquad (A1.15)$$

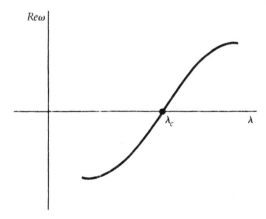

Figure A.1 Dependence of the real part of the stability exponent ω on the control parameter λ. The value $\lambda = \lambda_c$ marks the transition from asymptotic stability ($\lambda < \lambda_c$) to instability ($\lambda > \lambda_c$).

In many instances the rate functions v are space-independent. As a result, the entire spatial dependence of the solutions of the linearized equations [cf. Eqs. (A1.11)]

$$\frac{\partial x}{\partial t} = \left(\frac{\partial v}{\partial X}\right)_{X_s} \cdot x + D \cdot \nabla^2 x \tag{A1.16}$$

will be conditioned by the properties of the Laplace operator, ∇^2. The classical method for determining the possible forms of this spatial dependence, independently of all the complexities of the kinetics that are contained in the rate function v, is to solve the eigenvalue problem of the laplacian:

$$\nabla^2 \varphi_m(r) = -k_m^2 \varphi_m \tag{A1.17}$$

The eigenfunctions and eigenvalues $(\varphi_m, -k_m^2)$ will of course depend on the geometry, dimensionality, and boundary conditions. For instance, if Eq. (A1.17) is solved on a segment of length L with boundary conditions stipulating the vanishing of φ_m at $r = 0$ and $r = L$, we can easily verify that

$$k_m = \frac{m\pi}{L} \qquad \varphi_m(r) = \sin\frac{m\pi r}{L} \tag{A1.18}$$

where m is any nonvanishing integer. If, on the other hand, Eq. (A1.17) is solved on a two-dimensional circular layer, φ_m will be given by the products of Bessel functions and trigonometric functions.

Whatever the detailed form of φ_m might be, the solution of the linear equation (A1.11) will be [cf. Eq. (A1.12)]:

$$x = c\varphi_m(r)e^{\omega_m t} \tag{A1.19}$$

where c accounts for the difference between the state variables, and the eigenvalue ω of the linearized operator $\mathscr{L}$ is labeled by the same index m as the corresponding eigenfunction φ_m of the laplacian.

Substituting expression (A1.19) into Eq. (A1.16), we see that the operator $\partial/\partial t$ and ∇^2 act by giving, respectively, ω_m and $-k_m^2$ times x itself. As a result both φ_m and $e^{\omega_m t}$, which appear as common factors of all components of the vector x, cancel. The problem is thus reduced to

$$\omega_m c = \left(\frac{\partial v}{\partial X}\right)_{X_s} \cdot c - Dk_m^2 \cdot c$$

or, if D is diagonal,

$$(\omega_m + D_i k_m^2)c_i = \sum_j \left(\frac{\partial v_i}{\partial X_j}\right)_{X_s} c_j$$

This is a set of homogeneous algebraic equations for $\{c_i\}$. It admits a nontrivial solution provided that the determinant of the matrix of coefficients of c_i, vanishes. This gives rise to the solvability condition, or *characteristic equation*:

$$\det\left|\left(\frac{\partial v_i}{\partial X_j}\right)_{X_s} - \delta_{ij}^{kr}(D_i k_m{}^2 + \omega_m)\right| = 0 \qquad (A1.20)$$

where δ_{ij}^{kr} is the Kronecker symbol, $\delta_{ij}^{kr} = 0$ for $i \neq j$, $\delta_{ij}^{kr} = 1$ for $i = j$. This equation determines the eigenvalues ω_m in terms of the eigenvalues of the laplacian $k_m{}^2$, the diffusion coefficients D_i, and the other parameters contained in $(\partial v_i/\partial X_j)_{X_s}$.

A1.4 ILLUSTRATIONS

Let us illustrate the kind of information obtained from Eq. (A1.20) in a few representative examples.

One variable. A good example is the model of Eq. (2.15), whose rate equation is given by Eq. (A1.5). In the presence of diffusion the linearized operator is the sum of a scalar and the laplacian:

$$\mathscr{L} = (-3k_2 X_s{}^2 + 2k_1 a X_s - k_3) + D\nabla^2$$

As a result the characteristic equation reduces to

$$\omega_m = -3k_2 X_s{}^2 + 2k_1 a X_s - k_3 - Dk_m{}^2$$

and, more generally, for an arbitrary one-variable system:

$$\omega_m = \left(\frac{\partial v}{\partial X}\right)_{X_s} - Dk_m{}^2 \qquad (A1.21)$$

It follows that ω_m is real, hence the perturbations around the reference state evolve monotonically in time. Moreover, the contribution of the diffusion is negative or zero—in other words, diffusion cannot destabilize such a system. The first instability from X_s can therefore come only from the kinetics. This suggests that if a new solution past the instability exists, it will not display any intrinsic space dependence.

Two variables. A typical example is the brusselator model, for which the explicit form of the linearized operator $\mathscr{L}$ is given by Eq. (A1.8). The corresponding characteristic equation is given by the vanishing of the 2×2 determinant

$$\begin{vmatrix} B - 1 - (D_1 k_m^2 + \omega_m) & A^2 \\ -B & -A^2 - (D_2 k_m^2 + \omega_m) \end{vmatrix} = 0$$

or, in an explicit form,

$$\omega_m^2 - \omega_m[B - A^2 - 1 - (D_1 + D_2)k_m^2] + A^2$$
$$+ [-(B - 1)D_2 k_m^2 + A^2 D_1 k_m^2] + D_1 D_2 k_m^4 = 0 \quad (A1.22)$$

More generally, in a two-variable system the characteristic equation will be of the second degree in ω_m and will therefore possibly admit complex conjugate solutions. Setting $\omega_m = Re\omega_m + iIm\omega_m$, we find from Eq. (A1.20) that $Re\omega_m$ vanishes while $Im\omega_m$ remains finite if the following marginal stability condition is satisfied:

$$\left(\frac{\partial v_1}{\partial X_1}\right)_{X_s} + \left(\frac{\partial v_2}{\partial X_2}\right)_{X_s} = (D_1 + D_2)k_m^2 \quad (A1.23a)$$

For the brusselator this yields

$$B \equiv B_m = A^2 + 1 + (D_1 + D_2)k_m^2$$

and provides a relation between a parameter, here chosen to be B, and the eigenvalues of the laplacian k_m^2. For a general two-variable system a control parameter λ, contained in $(\partial v_i/\partial X_j)_{X_s}$, can always be chosen in such a way that Eq. (A1.23a) is represented by curve (a) of Figure A.2 (see also Figure 42). The corresponding imaginary part can be calculated from Eq. (A1.20) where Eq. (A1.23a) is substituted. In the brusselator this yields

$$\Omega_m^2 = (Im\omega_m)^2 = A^2 + [A^2 D_1 - (B_m - 1)D_2]k_m^2 + D_1 D_2 k_m^4$$

If new solutions of the nonlinear equations should exist beyond the marginal stability curve, we expect on these grounds that they will have a periodic behavior with an *intrinsic frequency* given essentially by Ω_m. On the other hand, they will still have a trivial space dependence, since the first unstable mode that will be excited corresponds again to $k_m = 0$.

The situation is very different if Eq. (A1.20) admits real ω_m's. The marginal stability condition becomes now simply $\omega_m = 0$, or

$$\det \left| \left(\frac{\partial v_i}{\partial X_j}\right)_{X_s} - \delta_{ij}^{kr} D_l k_m^2 \right| = 0 \quad (A1.23b)$$

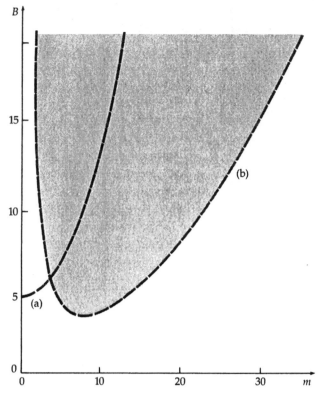

Figure A.2 Linear stability diagram for the brusselator in a one-dimensional medium. Bifurcation parameter B is plotted against wave number m. (a) and (b) are the marginal stability curves for two kinds of instability corresponding to complex conjugate and to real eigenvalues, respectively. Parameter values: $A = 2$; $L = 1$; $D_1 = 0.0016$; $D_2 = 0.008$. m and k_m are related through $k_m = m\,\pi/L$.

This relation is of fourth degree in k_m. If we plot a control parameter λ [which is contained in $(\partial v_i/\partial X_j)_{\mathbf{X}_s}$] against k_m, we may therefore expect an extremum at a certain nonvanishing k_{m_c} as indicated in curve (b) of Figure A.2 (cf. Figure 62). The first unstable mode that will be excited will thus be space-dependent. The point is that, in view of Eq. (A1.23b), k_{m_c} is entirely determined by the system's parameters, independent of size, geometry, or even dimensionality. Moreover, according to Eqs. (A1.19) and (A1.17), k_{m_c} determines the spatial wavelength of the disturbance acting on X_s, and for $\lambda \geqslant \lambda_c$ we expect this disturbance to determine the principal properties of the system past the instability. We therefore have a mechanism for the genesis of an *intrinsic space scale* in a hitherto spatially homogeneous system. We refer to this situation as *space symmetry-breaking*.

A1.5 SYSTEMS EXHIBITING CHAOTIC DYNAMICS

When the dynamics becomes chaotic, a stationary state on the thermodynamic branch ceases to be an appropriate reference state. Still, it is possible to characterize the instability of the motion and the resulting randomness of the behavior thanks to the *Lyapounov exponents*. Consider two phase space trajectories with initial conditions at two nearby points X_{0i} and $X_{0i} + \Delta x_{0i}$ respectively ($i = 1, \ldots, n$), as shown in Figure A.3. The evolution of their separation $\Delta x_i \equiv \Delta x_i(X_0, t)$ in the immediate vicinity of the point X_{0i} is essentially determined by the tangent vector of the "reference" trajectory, say ϕ, emanating from X_{0i}. It can thus be found by linearizing the equations for the variables X_i around their instantaneous values $X_i(t)$. By analogy to Section 3.6 [see Eq. (3.25)], we get

$$\frac{d\Delta x_i}{dt} = \sum_j L_{ij}(\{\phi_m(t)\})\, \Delta x_j \qquad (i = 1, \ldots, n) \tag{A1.24}$$

We now introduce the mean exponential rate of divergence of two initially close trajectories as follows:

$$\lambda_L(X_0, \Delta x_0) = \lim_{\substack{t \to \infty \\ |\Delta x_0| \to 0}} \frac{1}{t} \ln \frac{|\Delta x(X_0, t)|}{|\Delta x_0|} \tag{A1.25}$$

where Δx denotes the phase space vector whose components are $\{\Delta x_i\}$ and $|\Delta x|$ its norm. It can be shown that λ_L exists and is finite. Furthermore, as there are n independent directions in the phase space, when expressed in terms of projections along these directions λ_L takes n (possibly nondistinct) values, which are called the Lyapounov exponents. These are usually ordered by size,

$$\lambda_1 \geqslant \lambda_2 \geqslant \cdots \geqslant \lambda_n$$

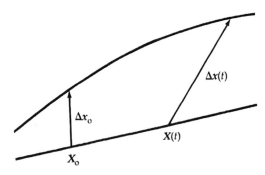

Figure A.3 Definition of Lyapounov characteristic exponents. Two initially nearby trajectories separate exponentially as time evolves.

and can be shown to be independent of the definition of the norm used. Note that for any flow generated by an autonomous system at least one Lyapounov exponent must vanish: in the direction along the flow $\Delta x(t)$ grows on the average at most linearly, and this yields in Eq. (A1.25) $\lambda_L \sim (1/t) \ln t \sim 0$ for large t. In a chaotic flow, on the other hand, the exponential divergence of the trajectories will be translated by the fact that at least one Lyapounov exponent is positive, since $|\Delta x(t)|/|\Delta x_0| \sim \exp(\lambda_\alpha t)$ and thus $1/t \ln |\Delta x(t)|/|\Delta x_0| \sim (1/t)\lambda_\alpha t \sim \lambda_\alpha > 0$.

The concept of Lyapounov exponent can also be used to describe the mean rate of expansion (or contraction) $\lambda^{(p)}$ of a phase space volume $\Delta V = \Delta x_{01} \cdots \Delta x_{0p}$, $p \leqslant N$. For $p = n$ this leads to

$$\lambda^{(n)} = \sum_{i=1}^{n} \lambda_i(X_0) \tag{A1.26}$$

In a conservative system the phase space volume remains invariant (cf. Section 3.2), hence $\lambda^{(n)} = 0$. In a dissipative system, on the other hand, the presence of attractors implies that there should be a global shrinking of the phase space volume, hence $\lambda^{(n)} < 0$. Thus, the presence of chaotic motions in conservative or dissipative systems necessarily implies that the phase space is expanding in certain directions and contracting in others. It is natural to conclude that in a chaotic motion variety and choice are continuously generated along the expanding directions of the motion, whereas predictability is recovered along the contracting directions. In other words, chaotic attractors (which, in addition to the above properties, enjoy asymptotic stability) are potential generators and processors of information. This point is discussed more fully in Section 4.8.

Bifurcation Analysis

A2.1 GENERAL PROPERTIES

The importance of linear stability analysis is to show that a qualitative change in the behavior of the system is bound to occur beyond the critical value λ_c of the control parameter. However, the existence of new physically acceptable solutions emerging beyond the threshold of instability will not be established until the full, nonlinear equations have been analyzed. In this appendix we outline some basic steps of this analysis.

The starting point is given by Eqs. (A1.10),

$$\frac{\partial x}{\partial t} = \mathscr{L}(\lambda) \cdot x + h(x, \lambda) \tag{A2.1}$$

where h stands for the nonlinear part of the righthand side of Eq. (A1.4). Before we describe the construction of the bifurcating solutions, it will be useful to comment on some general properties concerning existence and stability.

In Chapter 3 we produced a number of examples [cf. Eqs. (3.10), (3.12), (3.16)] showing that in certain cases it is indeed possible to establish the existence of new branches of solutions emerging from the parameter values at which the reference state loses its stability. However, we should refrain from inferring that this is always so.

To convince ourselves that more complex situations can arise, consider the nonlinear system

$$\frac{dX_1}{dt} = X_1 + X_2{}^3 - \lambda X_1 \qquad \frac{dX_2}{dt} = X_2 - X_1{}^3 - \lambda X_2$$

The reference state is $X_{1s} = X_{2s} = 0$, and the linearized problem around it has the form:

$$\omega x_1 = (1 - \lambda)x_1$$

$$\omega x_2 = (1 - \lambda)x_2$$

This linear system of homogeneous algebraic equations has a nontrivial solution only if $\omega = 1 - \lambda$. At $\lambda = 1$ we have a state of marginal stability. Actually, in the model under consideration $1 - \lambda$ is a double root of the characteristic equation, since the two equations of the linear approximation become uncoupled. The question we raise now is whether at the critical value $\lambda = 1$ of the parameter for which the solution of the characteristic equation is $\omega = 0$, we observe bifurcations of new branches of steady-state solutions. To see this we multiply the first equation by X_2, the second by X_1, and subtract. We obtain $X_1{}^4 + X_2{}^4 = 0$, which obviously has no real-valued (and thus physically acceptable) solution other than the trivial one. In other words, despite the fact that the linearized problem satisfies all requirements explained in Appendix 1, there exist no bifurcating steady-state branches.

If we look closely at the difference between the models considered in Sections 3.5 and 3.6 and the above model, we realize that in the former, the root of the characteristic equation, at least at the point of marginal stability, is simple, whereas in the latter it is always double. This is not a fortuitous correlation, but rather an illustration of the following general theorem:

If an eigenvalue of the linearized equations around a branch X_s has *odd multiplicity for* $\lambda \simeq \lambda_c$ (that is, if ω is a multiple root of the characteristic equation with odd multiplicity) and becomes zero at $\lambda = \lambda_c$, then there is at least one new branch of solutions outgoing from (X_s, λ_c). Moreover, this branch either extends to infinity or meets another bifurcation point.

We can obtain considerably more extensive information in the case of a simple eigenvalue and under the assumption that at λ_c not only does $Re\omega$ vanish, but the branch $Re\omega$ as a function of λ crosses the λ axis. This *transversality condition,*

$$\frac{d}{d\lambda} Re\omega(\lambda)\Big|_{\lambda = \lambda_c} \neq 0 \tag{A2.2}$$

is illustrated in Figure A.1. The following result can then be established:

If a branch (X_s, λ) asymptotically stable for $\lambda < \lambda_c$ loses its stability through a simple eigenvalue at $\lambda = \lambda_c$ and the transversality condition is satisfied at this critical value, then:

1. Supercritical bifurcating branches are stable, and subcritical bifurcating branches are unstable.

2. The bifurcating solutions from a stationary state will be stationary if $Im\omega(\lambda_c) = 0$, and time-periodic if $Im\omega(\lambda_c) \neq 0$.

Figure A.4 summarizes the various possibilities. We do not comment on them further, since they have all been illustrated amply throughout Chapter 3 and Appendix 1.

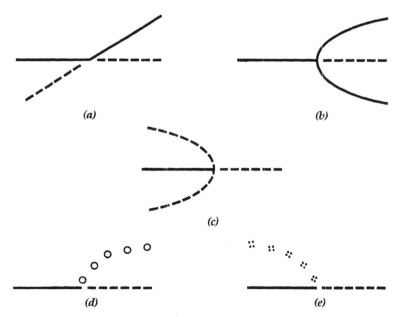

Figure A.4 Bifurcation diagrams around a steady state in the case of a simple eigenvalue of the linear stability operator. (a) Transcritical bifurcation of steady-state solutions, Eq. (A2.12). (b), (c) Supercritical and subcritical pitchfork bifurcation of steady state solutions, Eq. (A2.13). (d) Supercritical Hopf bifurcation. (e) Subcritical Hopf bifurcation.

A.2.2 EXPANSION OF THE SOLUTIONS IN PERTURBATION SERIES

We now assume that the conditions of validity of the above general theorems are satisfied, and outline the procedure for constructing the explicit form of the bifurcating branches. To simplify the technicalities we limit ourselves to bifurcation of steady-state solutions at a simple eigenvalue $[Im\omega(\lambda_c) = 0]$ arising as the result of the first instability of the reference state. The problem to be solved is then [cf. Eq. (A2.1)]:

$$\mathscr{L}(\lambda) \cdot x + h(x, \lambda) = 0 \qquad (A2.3)$$

We have repeatedly stressed the difficulties arising in the solution of nonlinear problems. For this reason we give up the idea of obtaining exact results of a global character, and limit our attention to the *local behavior* of the solutions in the vicinity of the bifurcation point λ_c. Furthermore, we suppose that the new solutions emerge at λ_c in a continuous fashion, thus excluding vertical branchings or jumps.

These hypotheses allow us to expand x in the vicinity of λ_c in power series of a small parameter. The latter must certainly be related to $\lambda - \lambda_c$, since at $\lambda = \lambda_c$ the norm $|x|$ of the solution goes to zero. There is no reason however that this small parameter be $\lambda - \lambda_c$ itself, since nothing guarantees in principle an analytic character for the solutions in $\lambda - \lambda_c$. In fact, in Chapter 3 we encountered several examples of manifestly nonanalytic dependence [cf., e.g., Eq. (3.11b)].

To allow the system itself to fix the dependence of the solutions on $\lambda - \lambda_c$ we introduce an auxiliary smallness parameter ε with respect to which we expand both x and $\lambda - \lambda_c$:

$$\begin{aligned} x &= \varepsilon x_1 + \varepsilon^2 x_2 + \cdots \\ \lambda - \lambda_c &= \varepsilon \gamma_1 + \varepsilon^2 \gamma_2 + \cdots \end{aligned} \qquad (A2.4)$$

Substituting into Eqs. (A2.3), we get, to the first order in ε,

$$\mathscr{L}(\lambda_c) \cdot x_1 = 0 \qquad (A2.5)$$

This homogeneous problem was discussed in detail in Appendix 1 in connection with linear stability analysis. Its solution is of the form [see Eq. (A1.12) with $\omega = 0$, the marginal stability value at $\lambda = \lambda_c$]:

$$x_1 = cu \qquad (A2.6)$$

For later use we have factored out the amplitude c of the solution, which is undetermined at this stage in view of the fact that Eqs. (A2.5) are homogeneous. It is therefore understood from now on that u is completely determined from the linearized problem. For instance, for the brusselator model we have from Eq. (A1.8):

$$(B_c - 1)u_1 + A^2u_2 - D_1k_c^2u_1 = 0$$
$$- B_cu_1 - A^2u_2 - D_2k_c^2u_2 = 0$$

from which the vector u can be simply computed as

$$u = \begin{bmatrix} 1 \\ - B_c/(A^2 + D_2k_c^2) \end{bmatrix}$$

We proceed now to higher orders in the perturbation analysis, which should allow us, in particular, to fix the amplitude c of the dominant part of the solution, εx_1. To order ε^2 we obtain from Eqs. (A2.3) three kinds of contributions. First, the operator $\mathscr{L}(\lambda_c)$ may act on the second-order term x_2 of the expansion of x. Second, we may evaluate the operator $\mathscr{L}(\lambda)$ at a λ close to λ_c, and have it act on x. This shift will be expressed formally by the first term of a Taylor expansion of $\mathscr{L}(\lambda)$ around λ_c and will thus yield the contribution $\gamma_1\mathscr{L}_\lambda(\lambda_c) \cdot x_1$, where derivatives are denoted by subscripts. As an example, if B is used as bifurcation parameter in the brusselator model, we see from Eqs. (A1.8) that

$$\mathscr{L}_\lambda(\lambda_c) \cdot x_1|_{\text{brusselator}} = \begin{pmatrix} x_1 \\ -x_1 \end{pmatrix}$$

Finally, to order ε^2 we will have a contribution coming from the quadratic part of h, which will again be denoted formally as the second term of a Taylor expansion around zero. For the brusselator model this will be simply the quadratic part of Eq. (A1.6a) at $B = B_c$:

$$\begin{bmatrix} (B_c/A)x^2 + 2Axy \\ (-B_c/A)x^2 - 2Axy \end{bmatrix}$$

Again denoting derivatives by subscripts, we finally obtain the full equation to order ε^2:

$$\mathscr{L}(\lambda_c) \cdot x_2 = -\gamma_1\mathscr{L}_\lambda(\lambda_c) \cdot x_1 - \tfrac{1}{2}h_{xx}(\lambda_c) \cdot x_1x_1 \qquad (A2.7)$$

Proceeding in exactly the same manner, we obtain to higher orders:

$$\mathscr{L}(\lambda_c) \cdot x_3 = -\gamma_1\mathscr{L}_\lambda(\lambda_c) \cdot x_2 - \gamma_2\mathscr{L}_\lambda(\lambda_c) \cdot x_1$$
$$- \frac{\gamma_1}{2} h_{xx\lambda}(\lambda_c) \cdot x_1x_1 - \frac{\gamma_1^2}{2} \mathscr{L}_{\lambda\lambda}(\lambda_c) \cdot x_1 \qquad (A2.8)$$
$$- \tfrac{1}{6}h_{xxx}(\lambda_c) \cdot x_1x_1x_1 - h_{xx}(\lambda_c) \cdot x_1x_2$$

and so on.

A2.3 *THE BIFURCATION EQUATIONS*

We have seen that, owing to the perturbation expansion, the initial nonlinear
problem [Eqs. (A2.3)] has been replaced by an infinite sequence of *linear* problems.
The first, Eq. (A2.5), is both linear and homogeneous, and its solution up to a
multiplicative factor is given by Eq. (A2.6). The remaining problems are inhomogeneous, as their righthand sides depend on the solution of the equations of the
lower orders. Remembering that $\mathscr{L}(\lambda_c)$ has the structure of a matrix in the space
of the vectors $\{x_i\}$, we are tempted to write, formally, their solutions as

$$x_i \quad \sim \mathscr{L}^{-1}(\lambda_c) \cdot q_i \qquad i \geqslant 2 \tag{A2.9}$$

where q_i denotes the righthand sides of Eqs. (A2.7, A2.8, etc.). The point however is that $\mathscr{L}(\lambda_c)$ is *not* invertible, since by Eq. (A2.5) there exist nontrivial
solutions on which its action gives zero. An analogous situation is encountered
in algebra, when in a set of linear inhomogeneous equations the determinant of
the coefficients of the unknowns vanishes. Intuitively, we expect that in such
cases the action of the inverse operator according to Eq. (A2.9) would produce
divergent results, unless perhaps the parts of q_i belonging to the null space of
$\mathscr{L}(\lambda_c)$ [the space of functions satisfying Eq. (A2.5)] could be eliminated. An important result of analysis, known as the theorem of the *Fredholm alternative*,
prescribes how this can be achieved. In brief, it stipulates that the righthand side q_i
should be *orthogonal* to the vectors u^+ that satisfy the equation

$$\mathscr{L}^+(\lambda_c) \cdot u^+ = 0 \tag{A2.10}$$

$\mathscr{L}^+$ being the adjoint operator of $\mathscr{L}$. For reaction-diffusion systems $\mathscr{L}^+$ is simply
the transpose matrix of $\mathscr{L}$, since the laplacian ∇^2 is self-adjoint for symmetric
boundary conditions. For instance, for the brusselator model:

$$\mathscr{L}^+(\lambda_c) = \begin{pmatrix} B_c - 1 + D_1\nabla^2 & -B_c \\ A^2 & -A^2 + D_2\nabla^2 \end{pmatrix}$$

How should orthogonality be understood in the above theorem? We have already insisted on the dual character of the state vector x, both as a vector in
an ordinary, finite-dimensional space and as a function in an infinitely dimensional
space. In ordinary finite-dimensional space orthogonality of two vectors is expressed by the vanishing of their scalar product, which we denote by a dot: $u^+ \cdot q_i$.
The scalar product in a function space is somewhat subtler and will not be explained here in detail. Suffice it to say that in many problems arising in quantum
mechanics and other areas of physics, it reduces to an integral of a product of
functions (or of a scalar product of vectors) over space. Denoting this generalized

product by brackets $\langle \ \rangle$, we can finally write the Fredholm alternative in the form

$$\langle u^+, q_i \rangle = 0 \qquad i \geqslant 2 \qquad (A2.11)$$

The point is that just like u, the vector u^+ is completely determined. Moreover, aside from its dependence on space coordinates, q_i depends only on the undetermined amplitude c, therefore, Eq. (A2.11)—in which it is understood that an integral over space is to be performed—reduces to an algebraic equation for the amplitude c of the dominant contribution to the bifurcating solution, x_1. Introducing the full amplitude $z = \varepsilon c$ of εx_1, we find from Eq. (A2.11) by straightforward algebra the following *bifurcation equations*:

$$-(\lambda - \lambda_c)Q_1 z + Q_2 z^2 = 0 \qquad \text{for } Q_2 \neq 0 \qquad (A2.12)$$

$$-(\lambda - \lambda_c)Q_1 z + Q_3 z^3 = 0 \qquad \text{for } Q_2 = 0, \ Q_3 \neq 0 \qquad (A2.13)$$

where the coefficients Q_i follow straightforwardly from our analysis. Their explicit values depend on the specific structure of the system under consideration.

Remember that we started from a problem involving a great number of variables that may depend in a complex way on the space coordinates. Yet in the vicinity of the bifurcation point it has been possible to reduce the dynamics to very simple, universal *normal forms*, similar to those discussed in Sections 3.4–3.5, involving only one unknown quantity, the *order parameter* z. This substantiates the arguments developed in the second part of Section 3.6.

More intricate bifurcation equations arise when time-periodic solutions or bifurcations near multiple eigenvalues are considered. For a succinct review of this topic we refer the reader to the book by Guckenheimer and Holmes listed in the Appendixes section of the suggested readings.

Perturbation of Resonant Motions in Nonintegrable Conservative Systems

A3.1 THE TWIST MAP

In Section 3.3 we saw that the motion of an integrable conservative system with N degrees of freedom lies on an N-dimensional toroidal surface embedded in the $(2N-1)$-dimensional phase space (cf. Figure 39 for $N = 2$). We want to understand intuitively the effect of nonintegrable perturbations on this kind of motion and, in particular, the qualitative changes induced near resonances that were anticipated in Section 3.8. For easier visualization we limit ourselves in this appendix to the case of two degrees of freedom, but we stress that the main arguments hold for more general systems as well.

Let θ, φ be respectively the two angles determining the "longitude" and the "latitude" on the torus shown in Figure A.5. ω_1 and ω_2 are the angular frequencies of the motion along these two directions. In angle-action variables the equations of motion are simply

$$I = I_0 = \text{const}$$
$$\varphi = \omega_1(I)t + \varphi_0 \qquad\qquad (\text{A3.1})$$
$$\theta = \omega_2(I)t + \theta_0$$

We have displayed only one action variable, I, since, by virtue of energy conservation, the second is expressed entirely in terms of I and the total energy. Moreover, we have allowed for the dependence of the angular frequencies on the action variable.

It is convenient to visualize the motion on a Poincaré surface of section S (see also Figure 54) along a meridional plane. The representative trajectories will intersect this surface every time that they perform a full turn along the θ direction, i.e., for $\theta - \theta_0 = 2\pi n$ $(n = 1, 2, \ldots)$. The intersections will therefore occur at times $t_n = 2\pi n/\omega_2$. Substituting this value of t in the equation for φ we obtain from Eqs. (A3.1):

$$\varphi_n - \varphi_0 = 2\pi \frac{\omega_1}{\omega_2} n$$

or, introducing notation appropriate for a recurrence relation,

$$I_{n+1} = I_n$$
$$\varphi_{n+1} = \varphi_n + 2\pi\alpha(I_n) \qquad\qquad (\text{A3.2})$$

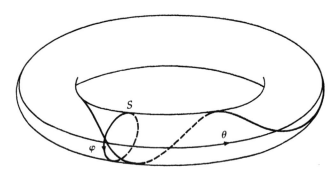

Figure A.5 Poincaré surface of section S suitable for studying resonant motions on a torus.

where the rotation number α is defined by

$$\alpha = \frac{\omega_1}{\omega_2} \qquad (A3.3)$$

The mapping given by Eqs. (A3.2) obviously maps circles onto circles. However, the rotation number will generally change from one circle to the other since it depends on the action variable, whose value labels the different circles. In other words, starting with a given value φ_0 of φ at $t = 0$, the representative points on different circles will be phase-shifted with respect to each other at some later intersection time $t = t_1$ (see, Figure A.6). For this reason the mapping, Eq. (A3.2), is referred to as a *twist mapping*. As pointed out in Section 3.8 in connection with Figure 54, each periodic trajectory on the torus ($\alpha = r/s$ rational, r and s being integers) cuts the Poincaré surface at fixed points whose number is equal to the number s of turns along the direction θ.

Consider now a slight perturbation of the underlying hamiltonian system. This will be reflected by a perturbed twist mapping of the form:

$$I_{n+1} = I_n + \varepsilon f(I_n, \varphi_n)$$
$$\varphi_{n+1} = \varphi_n + 2\pi\alpha(I_n) + \varepsilon g(I_n, \varphi_n) \qquad (\varepsilon \ll 1) \qquad (A3.4)$$

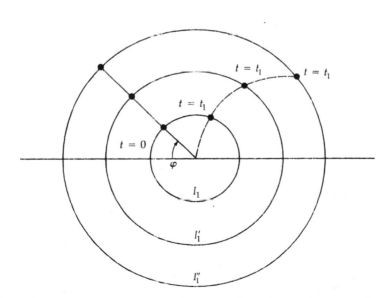

Figure A.6 The twist map. Images of three initial conditions lying on the same radius at $t = 0$ become phase shifted at $t = t_1$.

It can be shown that as a consequence of Liouville's theorem (Section 3.2), this mapping is bound to preserve the area in the phase space.

A3.2 EFFECT OF THE PERTURBATION IN THE CASE OF RATIONAL ROTATION NUMBERS

For the unperturbed twist mapping, Eqs. (A3.2), the above discussion shows that for $\alpha(I) = r/s =$ rational *any* point on the circle $I = const$ is a fixed point of period s, since it is bound to be the repeated intersection of the Poincaré surface with a certain periodic trajectory after the latter has performed s turns along the longitudinal direction. The major result concerning the effect of the perturbation on this picture goes back to Poincaré and Birkhoff. It states that out of this continuum of fixed points, $2ks$ fixed points remain after the perturbation (where k is an integer), half of which are hyperbolic and the other half elliptic. Let us outline the proof by referring to Figures A.7(a)–(d).

Suppose that the direction along which α is increasing is the inward one. Then the integrable system will necessarily possess two quasi-periodic trajectories, one outside and one inside the resonant torus. After s turns these trajectories will be mapped on the Poincaré surface shifted clockwise and counterclockwise, respectively, compared to the positions of their initial crossings [Figure A.7(a)].

Since quasi-periodic motion remains robust for small perturbations, there will still be two trajectories of the perturbed system having the above properties after s iterations of the perturbed twist mapping. Between the invariant curves generated on the Poincaré surface by these trajectories there must therefore be a curve, denoted by S_ε in Figure A.7(b) and enclosing the origin, which is constituted by the intersection points of those trajectories whose angular coordinate φ remains unchanged after s iterations of the perturbed twist mapping. Still, changes in the *radial* direction remain possible. Typically, therefore, after s iterations the curve S_ε will be displaced. We denote by S_ε' the image of S_ε generated in this way [Figure A.7(c)], which again encloses the origin. Because of its conservative character, the transformation preserves the area in the Poincaré surface. It follows that curves S_ε' and S_ε must enclose an equal area. But this is possible only if they intersect each other an even number of times. By construction, each of these intersection points returns to its initial position when iterated s times and is therefore a fixed point of period s. Each of its s iterates, which together would constitute the period s, will naturally be also a fixed point. In this way we have established the existence of fixed points enclosed between two nested invariant curves on the Poincaré surface, whose number is even and a multiple of s. Anticipating the result immediately to be derived, that elliptic fixed points alternate with hyperbolic ones, and using the obvious fact that an elliptic fixed point cannot be mapped on a hyperbolic one and vice versa, we arrive at the result $2ks$ for the

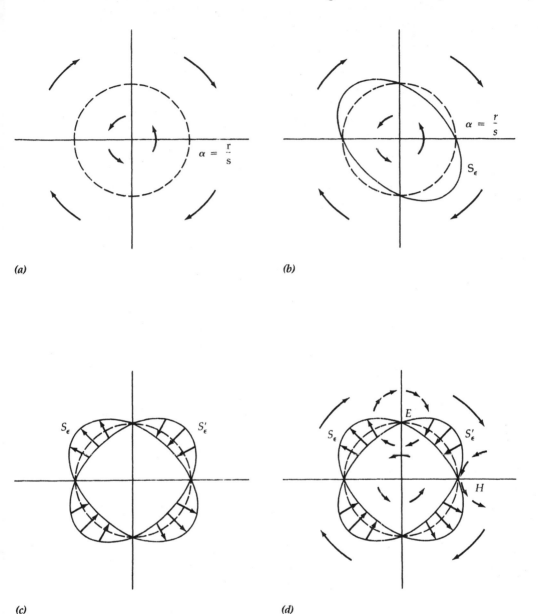

(a)

(b)

(c)

(d)

Figure A.7 Illustration of the Poincaré-Birkhoff theorem that some fixed points on the Poincaré surface S are preserved when a resonant motion is slightly perturbed. (a) Arrows on two sides of the dotted curve indicate the direction of the phase shifts experienced by two quasi-periodic trajectories in successive iterations of the mapping. (b) Locus of intersection of trajectories of the perturbed system that return to the same angular co-ordinate φ after s iterations, with Poincaré surface (curve S_ε). s is the number of turns along the angle θ. (c) Image of S_ε (curve S_ε') after s iterations of the perturbed mapping. Note the presence of an even number of fixed points. (d) Stability of the fixed points of the perturbed twist mapping. E, elliptic points; H, hyperbolic points.

number of fixed points that survive. The reasoning followed gives no information on the magnitude of the integer k.

We now discuss the stability of these fixed points. Consider the vicinity of point E in Figure A.7(d). A perturbation out of the fixed point, say in the downward direction, will essentially lead first to a clockwise rotation arising from the radial change of the rotation number α, then an inward motion (follow the arrows in the figure), then to a counterclockwise rotation, and finally to an outward motion. The overall result is a rotation around E, which behaves therefore as an elliptic (stable) fixed point. On the other hand, a perturbation away from the fixed point H shown in Figure A.7(d) leads the representative point farther and farther from H, which therefore behaves as a hyperbolic (unstable) fixed point. This alternation of fixed points of different stability provides a qualitative explanation of the main features of the phase portrait shown in Figure 55 of the text.

A3.3 HOMOCLINIC POINTS

What happens when an unstable fixed point of the hyperbolic kind (a saddle point) appears in the Poincaré map? In analogy to what we have found in Sections 3.3 to 3.5, in the vicinity of any such point there exist two exceptional curves, the separatrices, which join at this very point (Figure A.8, curves S and U). This remains compatible with the uniqueness of the solutions of the underlying dynamical system (see Section 3.1) as long as any initial condition on S or U cannot attain the fixed point H in finite time. Thus, if we consider an initial condition A_0 on the converging separatrix S, its first iterate by the mapping will lead to a state

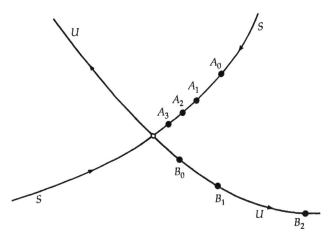

Figure A.8 Successive iterates of points A_0 and B_0 lying, respectively, on the stable and unstable separatrix of a hyperbolic fixed point.

A_1 on S closer to H, but the sequence A_0, A_1, ..., A_n will converge to H only for $n \rightarrow \infty$. Similarly, an initial condition B_0 on the diverging separatrix U will give rise for positive times to a set of states B_1, B_2, ... more and more removed from H, while for negative times it will give rise to a sequence similar to A_n, converging to H as $n \rightarrow -\infty$. The question we raise now is what happens to the separatrices far away from the fixed point H, especially when H coexists with many other hyperbolic fixed points?

Figure A.9 represents three possibilities: In the first case the separatrices join two of the coexisting hyperbolic points in such a way that the stable separatrix of one point becomes the unstable separatrix of the other, and vice versa. Such *heteroclinic curves* appear in integrable hamiltonian systems like the pendulum (see Figure 40). In the second case the two curves S and U become identified, and this results in a *homoclinic curve*. We alluded to this possibility in Section 3.10 (see Figure 61). A third possibility, depicted in Figure A.9(c), is simply that the unstable separatrix goes to infinity and the stable one comes from infinity.

The first two situations are rather exceptional, whereas the third one is physically unrealistic since it would imply (in the absence of other attracting fixed points) the explosive behavior of a certain variable describing the state of the system. Figure A.10 describes a much more natural situation, discovered by Poincaré: the unstable and stable separatrices of the fixed point H intersect at a point P, to which Poincaré gave the name *homoclinic point*. We again emphasize that such an intersection is not in contradiction with the uniqueness of the solutions of the underlying continuous dynamical system, even though P is not a fixed point of the map, since the invariant sets U and S of the mapping do *not* represent phase space trajectories of this system.

Now, Figure A.10 hides a rich behavior of an incredible complexity. As P belongs to the stable separatrix S, its iterates P_1, P_2, ..., P_n for positive times will all lie on S and will converge to the fixed point H as $n \rightarrow \infty$. On the other hand, P is also contained in the unstable separatrix U, and is therefore the image of certain points of this separatrix as the system is evolving on the Poincaré map by successive iterations. Alternatively, these points can be thought of as the

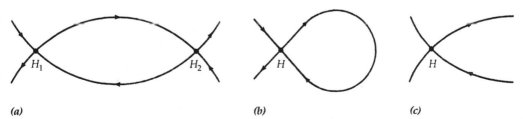

(a)   (b)   (c)

Figure A.9 Possible configurations of separatrices of hyperbolic fixed points. (a) Heteroclinic connection. (b) Homoclinic connection (separatrix loop). (c) Separatrices extend to infinity.

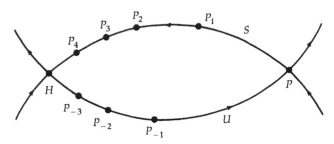

Figure A.10 Homoclinic point P, in a Poincaré surface of section: stable and unstable separatrices intersect transversally in P. P_1, P_2, ... represent successive iterates of P for positive times and lie on stable manifold S, whereas the preimages P_{-1}, P_{-2}, ... of P and lie on unstable manifold U.

iterates P_{-1}, P_{-2}, ..., P_{-n} of P for negative times, converging to the fixed point H as $n \to \infty$. But since P is later on rejected on P_1, P_2, ... the points P_{-1}, P_{-2}, ... will be mapped sooner or later on P_1, P_2, ··· as well. We thus reach the remarkable conclusion that curve U goes through the points P_1, P_2, ... of the curve S. In other words, P_1, ..., P_n, ... are also homoclinic points and so of

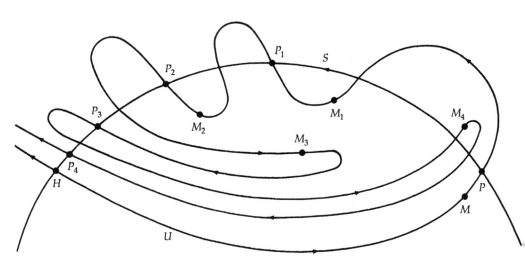

Figure A.11 Effect of a homoclinic point on the generation of complex motion near a separatrix. Stable and unstable manifolds of the hyperbolic point intersect infinitely many times. In conservative systems, area conservation implies that, in addition, these separatrices undulate strongly in the vicinity of H. After a sufficient number of iterations, a point M (not a preimage of P) initially on one side of the stable separatrix will be rejected on the other side (point M_4). One of the prerequisites of the onset of randomness is thus satisfied.

course are $P_{-1}, \ldots, P_{-n}, \ldots$. From a single homoclinic point P we have obtained a double infinity of such points!

It remains to see how the curves U and S will accommodate to this situation. Clearly they must bend and undulate continuously, with an increasing frequency near the hyperbolic fixed points. We can convince ourselves that these undulations are associated with ever larger excursions from the fixed point, followed by re-injections back to its vicinity, as indicated in Figure A.11. This explains qualitatively the statement made at the end of Section 3.9, namely, that the presence of homoclinic points gives rise to a dynamics similar to the horseshoe map. This suggests in turn the possibility of unstable motions of the chaotic type, in agreement with the results of numerical solution of the full nonlinear equations of evolution of nonintegrable conservative dynamical systems or of dissipative dynamical systems.

Reconstruction of the Dynamics of Complex Systems from Time Series Data. Application to Climatic Variability

A4.1 INTRODUCTORY COMMENTS

Experimentally, the behavior of a complex system is probed by observing a variable considered to be pertinent, during a certain period of time. An example of such a *time series* is provided by Figure 17, in which the changes of the global ice volume inferred from $^{18}O/^{16}O$ isotope data of deep sea core sediments are plotted against time. Another example is the electroencephalogram (EEG), the sum of elemental self-sustained neuronal activities of relatively low frequency (of the order of 0.5 to 40 hertz) emanating from small parts of cortical tissue close to the scalp.

At first sight, a time series of a single variable appears to provide a rather limited amount of information. In particular, it might be argued that such a series is restricted to a one-dimensional view of a system which, in reality, contains a large number of interdependent variables. In this appendix we show that a time series is far richer than that in information: it bears the marks of all other variables participating in the dynamics, and allows us to identify some key features of the underlying system independent of any modeling.

Let $X_0(t)$ be the time series derived from the experimental data. In actual fact of course, other variables $\{X_k(t)\}$, $k = 1, \ldots, n - 1$, are expected to take part in the dynamics. However, as we do not rely on any particular model, we want to reconstruct this dynamics solely from the knowledge of $X_0(t)$. To this end we consider the phase space spanned by the variables $\{X_k\}$, $k = 0, \ldots, n - 1$. As we saw in Chapter 3, an instantaneous state of the system becomes a point, say P, in this space, whereas a sequence of such states followed in time defines a curve, the phase space trajectory. As time progresses, the system reaches a state of permanent regime, provided its dynamics can be reduced to a set of dissipative deterministic laws. This is reflected by the convergence of families of phase trajectories toward a subset of the phase space. We have repeatedly referred to this invariant subset as the *attractor*. Our principal goal here is to find an answer to the following three questions:

1.   Is it possible to identify an attractor for a given time series? In other words, can the salient features of the system probed through this time series be viewed as the manifestation of a deterministic dynamics (possibly a very complex one), or do they contain an irreducible stochastic element?

2.   Provided that an attractor exists, what is its dimensionality d? We know from the analysis in Chapter 3 that the dimensionality provides us with valuable information on the system's dynamics. For instance, if $d = 1$ we are dealing with self-sustained *periodic* oscillations; if $d = 2$ we are in the presence of *quasi-periodic* oscillations of two incommensurate frequencies; and if d is a noninteger and larger than 2 (the case of a fractal attractor, Section 3.10), the system is expected to exhibit a chaotic oscillation featuring a great sensitivity to initial conditions, as well as an intrinsic unpredictability.

3.   What is the minimal dimensionality, n, of the phase space within which the above attractor is embedded? This defines the minimum number of variables that must be considered in the description of the underlying dynamics. Note that d is necessarily smaller than n.

Our first step is to identify a suitable set of variables spanning the phase space. A very convenient way is to unfold the original time series $X_0(t)$ by successively higher shifts defined as integer multiples of a fixed lag τ ($\tau = m \, \Delta t$, where m is an integer, and Δt is the interval between successive samplings). Taking, in addition, N equidistant points from the data set, we are led to the following set of discretized variables:

$$
\begin{aligned}
X_0: &\quad X_0(t_1), \ldots, X_0(t_N) \\
X_1: &\quad X_0(t_1 + \tau), \ldots, X_0(t_N + \tau) \\
&\vdots \\
X_{n-1}: &\quad X_0[t_1 + (n-1)\tau], \ldots, X_0[t_N + (n-1)\tau]
\end{aligned}
\tag{A4.1}
$$

For a proper choice of τ these variables are expected to be linearly independent, and this is all we need to define a phase space. Still, all these variables can be deduced from the single time series pertaining to $X_0(t)$, as provided by the data. We therefore see that, in principle, we have sufficient information at our disposal to go beyond the one-dimensional space of the original time series and to unfold the system's dynamics into a multidimensional phase space. This information allows us to draw the *phase portrait* of the system or, more precisely, its projection to a low-dimensional subspace of the full phase space. Figure A.12 depicts a three-dimensional phase portrait of the climatic evolution over the last million years

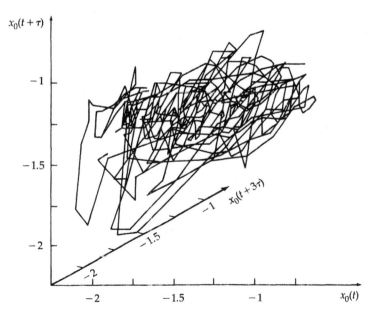

Figure A.12 Phase portrait representing the climatic evolution over the last one million years embedded in a three-dimensional space. The result utilizes about 500 equidistant values of X_0, the variable depicted in Figure 17. These are inferred from the oxygen isotope record obtained from the V28-238 deep sea core of Shackleton and Opdyke, following an interpolation available from the data bank of the University of Louvain. The value of the shift τ adopted in the figure is $\tau = 2000$ years.

inferred from the data of Figure 17. This view clearly exhibits the complexity of the underlying motion. However, the data are too coarse to draw any conclusion from this figure alone. Our next task is to characterize the complexity of the dynamics more sharply, using the techniques of the theory of dynamical systems.

A4.2 THEORETICAL BACKGROUND FOR DATA ANALYSIS

We introduce a vector notation: X_i stands for a point of phase space whose coordinates are $\{X_0(t_i), \ldots, X_0[t_i + (n-1)\tau]\}$. A reference point X_i is chosen from these data and all its distances $|X_i - X_j|$ from the $N - 1$ remaining points are computed. This allows us to count the data points that are within a prescribed distance, r, from point X_i in phase space. Repeating the process for all values of i, we arrive at the quantity

$$C(r) = \frac{1}{N^2} \sum_{\substack{i,j=1 \\ i \neq j}}^{N} \theta(r - |X_i - X_j|) \tag{A4.2}$$

where θ is the Heaviside function, $\theta(x) = 0$ if $x < 0$ and $\theta(x) = 1$ if $x > 0$. The nonvanishing of $C(r)$ measures the extent to which the presence of a data point X_i affects the position of the other points. $C(r)$ may thus be referred to as the integral *correlation function* of the attractor.

Suppose that we fix a given small ε and use it as a yardstick for probing the structure of the attractor. If the latter is a line, clearly the number of probing points within a distance r from a prescribed point should be proportional to r/ε. If it is a surface this number should be proportional to $(r/\varepsilon)^2$ and, more generally (cf. Section 3.7) if it is a d-dimensional manifold it should be proportional to $(r/\varepsilon)^d$. We therefore expect that for relatively small r, $C(r)$ should vary as

$$C(r) \sim r^d \tag{A4.3}$$

In other words, the dimensionality d of the attractor is given by the slope of the $\ln C(r)$ versus $\ln r$ in a certain range of r:

$$\ln C(r) = d \ln r \tag{A4.4}$$

The above results suggest the following algorithm:

1. Starting from a given time series construct the correlation function, Eq. (A4.2) by considering successively higher values of the dimensionality n of phase space.

2. Deduce the slope d near the origin according to Eq. (A4.4) and see how the result changes as n is increased.

3. If the d versus n dependence is saturated beyond some relatively small n, the system represented by the time series should possess an attractor. The saturation value d is regarded as the dimensionality of the attractor represented by the time series. The value of n beyond which saturation is observed provides the minimum number of variables necessary to model the behavior represented by the attractor.

A4.3 THE CLIMATIC ATTRACTOR

The above procedure has been applied to the analysis of the climatic data pertaining to Figure 17. Figure A.13 gives the dependence of ln $C(r)$ versus ln r for $n = 2$ to $n = 6$. We see that there is indeed an extended region over which this dependence is linear, in accordance with Eq. (A4.4). The circles in Figure A.14 show that the slope reaches a saturation value at $n = 4$, which is about $d \simeq 3.1$. The exes in the figure show how d varies with n if the signal considered corresponds to a gaussian white noise: there is no tendency to saturate. In fact, in this case d turns out to be equal to n. This establishes a clear-cut difference between

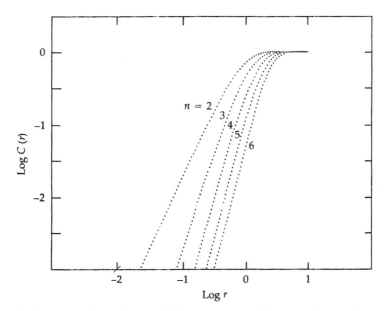

Figure A.13 Distance dependence of the correlation function of the climatic attractor. Note the existence of an extensive range of linearity, from which the dimensionality of the attractor can be inferred. Here $\tau = 8000$ years.

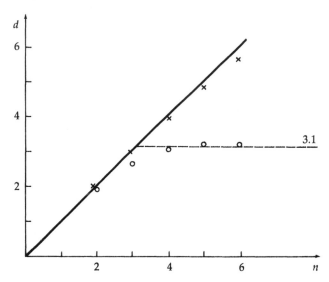

Figure A.14 Dependence of dimensionality d on the number of phase space variables n for the climatic attractor (∘) and for a white noise signal (×) for $\tau = 8000$ years. Notice the saturation to a plateau value of ~ 3.1 in the first case, and the $d \sim n$ relationship in the second case.

climate dynamics and random noise and provides strong evidence that climatic evolution over the last one million years can be viewed as the manifestation of a deterministic dynamics possessing a low-dimensional attractor. The estimation of the largest Lyapounov exponent can also be performed. One finds a positive exponent, corresponding to a predictability time of about 25 kyr. This provides us with a natural way to understand the intrinsic variability of the climatic system. A remarkable fact is that this dynamics is reducible to a limited set of key variables, although at this stage their specific nature cannot be identified.

It is instructive to confront this conclusion—which in a way is forced upon us by the data—with the results drawn from the models outlined briefly in Section 6.4. Two major points emerge. First, the instability of the motion on the chaotic attractor allows for the amplification of weak external signals, such as the small variations of the solar energy arising from the earth's orbital variability. The possibility of such an amplification was indeed one of the conclusions drawn in Section 6.4., although this phenomenon was attributed to the stochastic effects that continuously induced transitions between coexisting states.

The second point concerns the origin of the stochastic forcing that had to be added in the energy balance equation [see Eq. (6.8)]. As we saw in Sections 4.7 and 4.8, a chaotic attractor can generate a variety of stochastic processes. We therefore may regard the fluctuations in climate dynamics as one of the manifestations of the chaotic character of the underlying attractor.

A4.4 *CONCLUSION AND PERSPECTIVES*

The above approach has also been applied to EEG data corresponding to various stages of sleep. Again, the existence of an attractor for this system appears to be secured, at least for the deep sleep stage. Depending on the subject, its dimensionality turns out to be between about 4 and 4.3. On the other hand, in certain stages of pathological activity, such as epilepsy, one finds a much lower dimensionality, close to 2. In view of the comments made in Sections 4.7 and 4.8, it is tempting to relate this result to the very ability of the healthy brain to generate and process information. Other complex systems occurring in nature, such as short-term weather variability and economic activities, have also been analyzed from the same point of view.

We do not discuss these points in detail here, but refer the reader to the original publications, listed in the suggested readings.

Primordial Irreversible Processes

A5.1 INTRODUCTION

As Chap. 1 stressed, a theory of self-organization cannot be complete unless it takes into account gravitation, the most universal of all forces. The theory of general relativity forms the basis of the modern theory of gravitation. It has been called by Landau and Lifschitz "the most beautiful of all existing theories"; it certainly corresponds to a drastic revision of the newtonian description of our universe. Indeed, the newtonian view was based on a duality: on one side, space-time; on the other, matter (in fact, space and time were considered to be independent from each other as well as from the matter content). The basic novelty of Einstein's general relativity was to suppress this duality and to establish a connection between space-time on one side and matter on the other.

Einstein's field equations relate the properties of space-time to the properties of matter; the former are essentially related to curvature, while the latter are related to pressure and energy density. In this view, many space-times are possible: flat space-time or curved spaces. Space-time becomes itself a physical phenomenon, and we may enquire about its origin and its evolution. This is in fact one of the basic aims of modern cosmology. It is quite striking that Einstein, as early as 1917, one year after the formulation of general relativity, presented the first theoretical model of cosmology. Einstein wanted to arrive at a timeless description of the universe; for him, irreversibility and the arrow of time were expressions of subjectivity that had to be eliminated by science. In accordance with these views, Einstein presented a static model of the universe.

The subsequent evolution of cosmology is one of the most dramatic examples of the intrusion of time into modern physics. In 1923, Alexander Friedmann showed that Einstein's universe was unstable, and proposed the theory of an expanding universe, later validated by the observations of Edwin Hubble (1929): galaxies, as seen from the earth, recess with a speed proportional to their distance. The concept of an evolutionary universe was born. Still, one could have considered that this was a purely geometrical feature, of no importance for physics.

In 1965, however, Penzias and Wilson discovered the existence of the residual blackbody radiation: our universe is full of photons that are in a thermal equilibrium corresponding to a temperature of ~ 2.7 K (it is because this temperature is so low that it was so difficult to identify this radiation experimentally). This radiation was produced a long time ago, during the first stages of our universe (whence the term "residual"). It appeared as a result of cosmological expansion, when the temperature became low enough to permit atoms to be formed; at this point the radiation became decoupled from matter. Since the discovery of the residual blackbody radiation, we know that the universe has a dual structure: on one side, massive particles (baryons); on the other, massless particles (e.g., photons). In fact, baryons are largely outnumbered by photons: there are hundreds of millions photons for one baryon. Moreover, it is striking that the entropy related to this residual blackbody radiation makes up by far the largest part of the entropy of the universe. It has been estimated that if all of the matter of our universe would evaporate, this would only change the total entropy by a fraction of 1 percent. This has remarkable thermodynamic consequences. We have seen that the classical formulation of the second principle predicts thermal death (maximum entropy) in our future; but we understand now that at the start of our universe so much entropy was produced that we may say that thermal death lies in fact behind us. We have therefore to analyze the irreversible processes at the start of the universe more closely.

A5.2 STANDARD COSMOLOGICAL MODEL

As already mentioned, modern cosmology is based on Einstein's equations. Furthermore, it assumes that the universe is uniform and isotropic (this is the so-called cosmological principle). Given this limit, Einstein's equations take a very simple form. We shall write these equations without proof for the simplest case corresponding to zero pressure and euclidean space, in which only two variables are involved. One of them, sometimes referred to as the "radius" of the universe, R, determines the limits that we can reach through astronomical observation; the other is the energy density ρ. Einstein's equations relate ρ to radius R through

$$\kappa\rho = 3H^2 \tag{A5.1}$$

$$\dot{\rho} = -3H\rho \tag{A5.2}$$

where $H = \dot{R}/R$ is the Hubble function, and the dot denotes the derivative with respect to cosmological time. The second equation has an especially simple meaning: the evolution of the universe is adiabatic and reversible; Einstein's equations conserve entropy. More generally, we have three variables: the universe radius, the energy density ρ, and the pressure p. We need therefore an equation of state $p(\rho)$. We also have to take into account the space curvature.

For homogeneous and isotropic universes, there are only three possible cases, corresponding to negative, zero, or positive curvature. They give rise respectively to open space (negative spatial curvature), euclidean space (zero spatial curvature), and closed space (positive spatial curvature). All these solutions start with a universe concentrated in a single point. The standard model leads therefore to a primordial singularity, the big bang. This conclusion has been called by John A. Wheeler "the greatest crisis of physics." Indeed, what could be the meaning of such a singularity? If we retrace the history of the universe into the past, do we indeed come to a point beyond which no extrapolation is positive, a point beyond which the laws of physics cease to be valid? Do we actually have to deal with primordial singularity, or do we have to consider the start of the universe as resulting from some kind of instability, associated to a phase transition like phenomenon?

A5.3 BLACK HOLES

Whatever the answer, there exists a consensus among physicists that these events happened during the so-called Planck era, dominated by three fundamental constants: c (the speed of light), κ (the gravitational constant), and h (the Planck constant), out of which one can derive the Planck time ($\sim 10^{-44}$ s, a very short

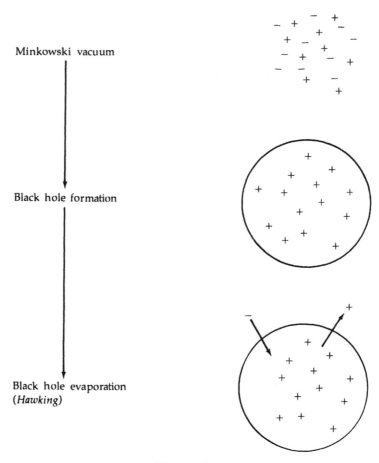

Minkowski vacuum

Black hole formation

Black hole evaporation
(*Hawking*)

Figure A.15

time indeed), the Planck mass ($\sim 10^{-5}$ g, very large compared to the mass of a proton), and the Planck length ($\sim 10^{-33}$ cm, which is very small). We still do not have a fundamental theory of Planck's era. This would indeed involve a quantum theory of gravitation, which is a subject facing great difficulties. But it is interesting that some aspects of quantum theory of gravitation can be obtained without solving the basic problems associated to quantum theory fields. An example is Stephen Hawking's theory of black holes, which leads to quite remarkable predictions (Figure A15).

A black hole can be thought of as a "membrane" enclosing a space-time region that does not permit light to escape. The boundaries of black holes are called the "event horizon": whenever light encounters this event horizon, it remains prisoner (whence the term "black" holes). But in fact, as Stephen Hawking and others have shown, black holes are not so "black": they do radiate when they are embedded into the vacuum, that is, in a state of space-time in which the density of matter and energy are vanishing. Moreover, they radiate like a black body with a temperature T, which depends only on the mass of the black hole. The origin of this thermal radiation comes from their interaction with the quantum mechanical vacuum. As we have mentioned in Section 5.10, quantum mechanical vacuum is the locus of fluctuations. These fluctuations lead to the formation of pairs of particles. But these pairs are virtual pairs; they exist only for a short time. One of the particles has a positive energy, the other a negative one. Hawking's radiation results from the fact that a black hole absorbs particles of negative energy. Therefore, we have a flow of positive energy going away from the black hole, and the black hole evaporates. Black holes lead therefore to irreversible processes.

A5.4 THE ROLE OF IRREVERSIBILITY

We now come to the question of whether the origin of our universe is a singularity or an instability. An idea often presented is that our universe would be a "free lunch", that is, it derives from the fact that all available energy is ultimately present under two fundamental forms, which compensate each other: mass-related energy (which is positive) and gravitation-related energy (which is negative). Thus, from the point of view of energy, there is no difference between the vacuum (the Minkowski universe) and a material universe.

We have discussed previously the Bénard instability. There also, the price to pay is not labeled in terms of energy; in the regime of stationary convection, the energy is of course conserved, and in this sense we also have a "free lunch." The entropy production makes the difference. It is therefore tempting to think that the price to pay for the birth of our universe is also entropy production. To explain this, we have to study more carefully the impact of creation of matter on the two principles of thermodynamics, as well as on Einstein's equations.

Let us consider a volume V containing N particles. For a closed system, N is constant. The conservation of internal energy E is then expressed by

$$dE = dQ - p\,dV \qquad (A5.3)$$

where dQ is the heat received by the system during a time dt. We may rewrite the relation in Eq. (A5.3) in the form

$$d\left(\frac{\rho}{n}\right) = dq - pd\left(\frac{1}{n}\right) \qquad \text{with} \qquad \rho = \frac{E}{V}, \quad n = \frac{N}{V}, \quad dq = \frac{dQ}{N} \qquad (A5.4)$$

The relation in Eq. (A5.4) being local, it remains valid for open systems in which N is time dependent. In this case, it leads to a modification of Eq. (A5.3) that takes explicitly into account the variation of the number of particles:

$$d(\rho V) = dQ - p\,dV + \frac{h}{n}\,d(nV) \tag{A5.5}$$

where $h = \rho + p$ is the enthalpy per unit volume. For closed systems, adiabatic transformations ($dQ = 0$) are defined by the relation

$$d(\rho V) + p\,dV = 0 \tag{A5.6}$$

The extension to open systems is given by the equation

$$d(\rho V) + p\,dV - \frac{h}{n}\,d(nV) = 0 \tag{A5.7}$$

In this transformation, the "heat" received by the system is entirely due to the change of the number of particles. In our cosmological perspective, this change is due to the transfer of energy from gravitation to matter. Hence, *the creation of matter acts as a source of internal energy.*

We turn now to the second law of thermodynamics (see Section 2.5). As usual, we decompose the entropy change into an entropy flow d_eS and an entropy creation d_iS:

$$dS = d_eS + d_iS \qquad \text{with} \qquad d_iS \geqslant 0 \tag{A5.8}$$

To evaluate the entropy flow and the entropy production, we start from the total differential of the entropy

$$T\,d(sV) = d(\rho V) + p\,dV - \mu\,d(nV) \tag{A5.9}$$

where $s = S/V$, μ is the chemical potential, $\mu n = h - Ts$, and $s \geqslant 0$. (A5.10)

For closed systems and adiabatic transformations the relation in Eq. (A5.9) leads to

$$dS = 0 \quad \text{and} \quad d_iS = 0 \tag{A5.11}$$

Let us consider the effect of matter creation. We consider homogeneous systems and therefore expect that we still have $d_eS = 0$. In contrast, we stipulate that matter creation contributes to entropy production. We have therefore:

$$T \, d_i S = T \, dS = \frac{h}{n} \, d(nV) - \mu \, d(nV) = T \frac{s}{n} \, d(nV) \geqslant 0 \qquad \text{(A5.12)}$$

This inequality in the cosmological context implies that space-time can produce matter, while the reverse process is thermodynamically forbidden. The relation between space-time and matter ceases to be symmetrical, since particle production, occurring at the expense of gravitational energy, appears to be an irreversible process. The microscopic interpretation of this process in which space-time is converted into matter will be briefly discussed at the end of this Appendix.

The relation in Eq. (A5.7) can be written in a number of equivalent forms such as:

$$\dot{\rho} = \frac{h}{n} \dot{n} \qquad \text{(A5.13)}$$

$$p = \frac{n\dot{\rho} - \rho\dot{n}}{\dot{n}} \qquad \text{(A5.14)}$$

It is interesting to note that energy creation $\dot{\rho}$ and particle creation $\dot{n}$ determine the pressure p. As example, let us note that $\rho = Mn$ implies $p = 0$ and furthermore, $\rho = \alpha T^4$, $n = bT^3$ implies $p = \rho/3$.

In order to discuss Einstein's equations for open systems, it is useful to transcribe the relation in Eq. (A5.7) in a form akin to that of Eq. (A5.6), namely

$$d(\rho V) = -(p + p_c) \, dV \qquad \text{(A5.15)}$$

where

$$p_c = -\frac{h}{n} \frac{d(nV)}{dV} = -\frac{\rho + p}{n} \frac{d(nV)}{dV} \qquad \text{(A5.16)}$$

corresponds to a supplementary pressure due to matter creation. Because of Eq. (A5.12), p_c is always negative. We may now indicate the change related to matter creation in Einstein's equations. As we have noted, Einstein's equations relate the properties of space-time (curvature) to density and to pressure. When there is matter creation, we have to introduce the corresponding pressure $\tilde{p}$ in Einstein's equations as the sum

$$\tilde{p} = p + p_c \qquad \text{(A5.17)}$$

of the usual pressure p plus the pressure p_c due to creation of matter.

We obtain in this way a class of new cosmologies, in which the variables are R, r, p, and n. Let us consider a simple case

$$\rho = Mn \tag{A5.18}$$

that implies $p = 0$ for open systems. We are left with two variables, R and n. One equation is provided by Eq. (A5.1). We need a supplementary equation, relating the creation of particles to the Hubble function H. We take for the volume V the value

$$V \sim R^3 \tag{A5.19}$$

The simplest kinetic equation satisfying the inequality in Eq. (A5.12) is

$$\frac{1}{R^3} \frac{dnR^3}{dt} = \alpha H^2 \geqslant 0 \tag{A5.20}$$

where $\alpha \geqslant 0$. For $\alpha = 0$ we recover the usual standard model description, including the big bang singularity. However, for $\alpha \neq 0$, the situation changes drastically. We may solve Eqs. (A5.1) and (A5.20), and show that whatever the initial values are, the system tends to a *de Sitter space*, characterized by n and H constant in time. This corresponds to an exponential expansion of the universe, as $H = \dot{R}/R$. This result is quite independent of the form of Eq. (A5.20). The introduction of matter creation thus makes the big bang structurally unstable, and the slightest creation of matter leads to an exponentially expanding universe. At a more microscopic level, the creation of matter out of vacuum may be regarded as an effect of the "instability" of vacuum in respect to its fluctuations. We have seen that quantum vacuum is associated with the appearance of virtual particles. If a large enough number of these particles appears in a given volume (as the result of fluctuations), this may lead to the temporary formation of a black hole, which would then decay, according to the Hawking mechanism. In this way, virtual particles are transformed into actual ones. We may therefore envisage a cosmological history in three stages: first, the instability of the vacuum from which emerges inevitably a de Sitter universe; next, the process of black hole evaporation, which is assumed to inhibit further creation; and finally, the switching from the de Sitter phase to the phase of our present, adiabatically expanding universe, as described by the standard model.

A qualitative treatment of this cosmology introduces as the basic parameter the mass of the black holes m_{bh} that act as intermediary between virtual and actual particles. For $m_{bh} \sim 100\ m_{Planck}$, we obtain an excellent agreement with the present value of the basic thermodynamic quantities of our universe, such as the ratio of baryons to photons. As the lifetime of black holes is very short, the two first stages correspond also to a very short time, of the order of 10^{-37} s, which is 10^7 longer than the Planck time.

It is interesting to notice that the quantum formulation of this problem gives a similar order of magnitude for the mass of the black holes during the process of formation of our universe. In terms of our model, we would have some kind of phase transition, separating matter from gravitation. These two components of our universe were intimately mixed in the primordial vacuum. Modern cosmology closely mixes microscopic and macroscopic aspects. We could not obtain this description of the early stages of the universe without Planck's constant ruling the microscopic world. This expresses beautifully what is often called the holistic character of the universe. Of course, what is at stake here is not creation from nothing: the Minkowski vacuum is already a well-defined medium, characterized by the values of the three fundamental constants: c (the speed of light), κ (the gravitational constant), and h (the Planck constant). One could of course ask more ambitious questions, such as where do these constants come from? And are they invariant in time or are they themselves changing? There is at present no theoretical guide to answer these questions.

At the beginning of this Appendix, we mentioned the duality of Newton's world. We have emphasized the unifying role of Einstein's general relativity. We may look at Einstein's equations from a standpoint similar to that of an engineer emphasizing the equivalence of heat and mechanical work: Einstein's equations express a kind of equivalence between space-time and matter. But as is the case for heat and work, this is not a full equivalence. Space-time appears as a "noble form" out of which matter would emerge through irreversible process involving a huge entropy production. One of the recurring themes of this book is that entropy production involves the production of both order and disorder, in agreement with the conspicuous duality of our universe.

Suggestions for Further Reading

GENERAL

Physical Aspects

Presentations of the physical ideas underlying irreversible thermodynamics, self-organization, and complexity can be found in the following sources.

Babloyantz, A. 1986. *Molecules, dynamics and life*. Wiley, New York.
Careri, G. 1984. *Order and disorder in matter*. Benjamin Cummings, Menlo Park, California.
Ebeling, W., and R. Feistel. 1982. *Physik der Selbstorganisation und Evolution*. Akademie-Verlag, Berlin.

Glansdorff, P., and I. Prigogine. 1971. *Thermodynamics of structure, stability and fluctuations.* Wiley, London.

Haken, H. 1977. *Synergetics.* Springer, Berlin.

Nicolis, J.S. 1986. *Dynamics of hierarchical systems.* Springer, Berlin.

Nicolis, G., and I. Prigogine. 1977. *Self-organization in nonequilibrium systems.* Wiley, New York.

Peacocke, A. 1983. *The physical chemistry of biological organization.* Clarendon, Oxford.

Prigogine, I. 1961. *Introduction to thermodynamics of irreversible processes.* Wiley, New York.

Prigogine, I. 1980. *From being to becoming.* Freeman, San Francisco.

Thompson, J.M.T. 1982. *Instabilities and catastrophes in science and engineering.* Wiley, Chichester.

Mathematical Aspects

Presentations of fractal geometry, the theory of nonlinear dynamical systems, and chaotic dynamics can be found in the following sources.

Arnold, V. 1980. *Chapitres supplémentaires de la théorie des équations différentielles ordinaires.* Mir, Moscow.

Bergé, P., Y. Pomeau, and C. Vidal. 1984. *L'ordre dans le chaos.* Herman, Paris.

Guckenheimer, J., and Ph. Holmes. 1983. *Nonlinear oscillations, dynamical systems and bifurcations of vector fields.* Springer, Berlin.

Lichtenberg, A., and M. Lieberman. 1983. *Regular and stochastic motion.* Springer, Berlin.

Mandelbrot, B. 1977. *Fractals: form, chance, dimension.* Freeman, San Francisco.

Sattinger, D. 1973. *Topics in stability and bifurcation theory.* Springer, Berlin.

Schuster, H. 1984. *Deterministic chaos.* Physik-Verlag, Weinheim.

CHAPTER 1

Thermal Convection

Chandrasekhar, S. 1961. *Hydrodynamic and hydromagnetic stability.* Oxford University Press, Oxford.

Koschmieder, E.L. 1981. In G. Nicolis, G. Dewel, and J.W. Turner (eds.). *Order and fluctuations in equilibrium and nonequilibrium statistical mechanics.* Wiley, New York.

Velarde, M., and C. Normant. 1980. *Scientific American* 243(1), 78.

Self-Organization Phenomena in Chemistry, Interfaces,
Materials Science, and Geology

Field, R., and M. Burger (eds.). 1985. *Oscillations and travelling waves in chemical systems.* Wiley, New York.
Müller, S., T. Plesser, and B. Hess. 1986. *Naturwissenschaften 73,* 165.
Nicolis, G., and F. Baras (eds.). 1984. *Chemical instabilities.* D. Reidel, Dordrecht.
Nicolis, G., and J. Portnow. 1973. *Chem. Rev. 73,* 365.
Pacault, A., and C. Vidal (eds.). 1978. *Far from equilibrium.* Springer, Berlin.
Vidal, C., and A. Pacault (eds.). 1981. *Nonlinear phenomena in chemical dynamics.* Springer, Berlin.
Vidal, C., and A. Pacault (eds.). 1984. *Nonequilibrium dynamics in chemical systems.* Springer, Berlin.
Winfree, A. 1972. *Science 175,* 634.
Zhabotinski, A. 1964. *Biofizika 9,* 306.

Quantum Optics

Abraham, E., and S. Smith. 1982. *Rep. Progr. Phys. 45,* 815.
Bonifacio, R., and L. Lugiato. 1976. *Opt. Commun. 19,* 172.
Gibbs, H., S. McCall, and T. Venkatesan. 1976. *Phys. Rev. Lett. 36,* 113.
Haken, H. 1977. *Synergetics.* Springer, Berlin.

Control of Developmental Processes

Goldbeter, A., and L. Segel. 1977. *Proc. Nat. Acad. Sci. U.S. 74,* 1543.
Kauffman, S., R. Shymko, and E. Trabert. 1978. *Science 199,* 259.
Segel, L. 1984. *Modeling dynamical phenomena in molecular and cellular Biology.* Cambridge University Press, Cambridge, England.
Sussman, M. 1964. *Growth and development.* Prentice-Hall, Englewood Cliffs, New Jersey.
Turing, A. 1952. *Phil. Trans. Roy. Soc. B237,* 37.

Climatic Variability

Berger, A. (ed.). 1981. *Climatic variations and variability: facts and theories.* Reidel, Dordrecht.
Lamb, H. 1977. *Climate: present, past and future.* Methuen, London.
Schneider, S. 1976. *The genesis strategy.* Plenum, New York.

Phase Transitions, Cosmology, and Symmetry Breaking

Georgi, H. 1981. *Pour la science 44,* 68.
Guth, A., and P. Steinhardt. 1984. L'univers inflatoire. *Pour la Science 81,* 86.

Hawking, S. 1988. *A brief history of time, from the big bang to black holes.* Bantam, New York.
Landau, L., and E. Lifshitz. 1980. *Statistical physics.* 3d ed. Pergamon, Oxford.
Ma, S. 1976. *Modern theory of critical phenomena.* Benjamin, Reading, Massachusetts.
Stanley, H. 1971. *Introduction to phase transitions and critical phenomena.* Clarendon, Oxford.
Weinberg, S. 1972. *Relativity and cosmology.* Wiley, New York.
Weinberg, S. 1977. *The first three minutes of the universe.* Basic Books, New York.

Algorithmic Complexity

Chaitin, G. 1975. *Scientific American* 233(5), 47.
Nicolis, J.S. 1986. *Rep. Progr. Phys.* 49, 1109.

CHAPTER 2

Newtonian and Hamiltonian Formalisms of Classical Mechanics

Goldstein, H. 1959. *Classical mechanics.* Addison-Wesley, Reading, Massachusetts.

Thermodynamic Description of Dissipative Systems

De Groot, S., and P. Mazur. 1962. *Nonequilibrium thermodynamics.* North-Holland, Amsterdam.
Katchalsky, A., and P. Curran. 1965. *Nonequilibrium thermodynamics in biophysics.* Harvard University Press, Cambridge, Massachusetts.
Onsager, L. 1931. *Phys. Rev.* 37, 405; 38, 2265.
Prigogine, I. 1961. *Introduction to thermodynamics of irreversible processes.* Wiley, New York.

Simple Models of Nonlinear Systems

Prigogine, I., and R. Lefever. 1968. *J. Chem. Phys.* 48, 1695.
Schlögl, F. 1972. *Z. Phys.* 253, 147.
Volterra, V. 1936. *Leçons sur la théorie mathématique de la lutte pour la vie.* Gauthier-Villars, Paris.

Basic Definitions of Stability and Bifurcation

Andronov, A., A. Vit, and C. Khaikin. 1966. *Theory of oscillators.* Pergamon, Oxford.

Poisson Distribution

Feller, W. 1959. *An introduction to probability theory and its applications,* vol. I. Wiley, New York.

CHAPTER 3

Phase Space Description, Liouville Equation

Gibbs, J.W. 1960. *Elementary principles in statistical mechanics.* Dover, New York.
Prigogine, I. 1962. *Nonequilibrium statistical mechanics.* Wiley, New York.

Stable and Unstable Motions in Conservative Systems

Arnold, V. 1976. *Methodes mathématiques de la mécanique classique.* Mir, Moscow.
Arnold, V., and A. Avez. 1968. *Ergodic problems of classical mechanics.* Benjamin, New York.
Birkhoff, G. 1927. *Dynamical systems.* AMS Publ., Providence, Rhode Island.
Kolmogorov, A. 1954. *Dokl. Akad. Nauk SSSR 98,* 527.
Lichtenberg, A., and M. Lieberman. 1983. *Regular and stochastic motion.* Springer, Berlin.
Moser, J. 1973. *Stable and random motions in dynamical systems.* Princeton University Press, Princeton, New Jersey.
Poincaré, H. 1899. *Les méthodes nouvelles de la mécanique céleste.* Gauthier-Villars, Paris.

Instabilities and Bifurcations in Dissipative Systems

Andronov, A., A. Vit, and C. Khaikin. 1966. *Theory of oscillators.* Pergamon, Oxford.
Arnold, V. 1980. *Chapitres supplémentaires de la théorie des équations différentielles ordinaires.* Mir, Moscow.
Guckenheimer, J., and Ph. Holmes. 1983. *Nonlinear oscillations, dynamical systems, and bifurcations of vector fields.* Springer, Berlin.
Haken, H. 1977. *Synergetics.* Springer, Berlin.

Helleman, R., and O. Rössler (eds.). 1979. *Bifurcation theory and applications in scientific disciplines.* Ann. New York Acad. Sci. 316.
Iooss, G., and D. Joseph. 1981. *Elementary stability and bifurcation theory.* Springer, Berlin.
Nicolis, G., and I. Prigogine. 1977. *Self-organization in nonequilibrium systems.* Wiley, New York.
Sattinger, D. 1973. *Topics in stability and bifurcation theory.* Springer, Berlin.
Thom, R. 1972. *Stabilité structurelle et morphogénèse.* Benjamin, Reading, Massachusetts.
Wasow, W. 1965. *Asymptotic expansions for ordinary differential equations.* Wiley-Interscience, New York.

Chaotic Dynamics, Fractals

Arneodo, A., P. Coullet, and C. Tresser. 1981. *Commun. Math. Phys. 79,* 573.
Arneodo, A., P. Coullet, and C. Tresser. 1982. *J. Stat. Phys. 27,* 171.
Bergé, P., Y. Pomeau, and C. Vidal. 1984. *L'ordre dans le chaos.* Herman, Paris.
Collet, P., and J.P. Eckmann. 1980. *Iterated maps on the interval as dynamical systems.* Birkhäuser, Basel.
Feigenbaum, M. 1978. *J. Stat. Phys. 19,* 25.
Ford, J. 1983. *Physics Today,* April, 40.
Gaspard, P., and G. Nicolis. 1983. *J. Stat. Phys. 31,* 499.
Gumowski, I., and C. Mira. 1980. *Recurrences and discrete dynamical systems.* Springer, Berlin.
Lorenz, E. 1963. *J. Atmos. Sci. 20,* 130.
Mandelbrot, B. 1977. *Fractals: form, chance, dimension.* Freeman, San Francisco.
Nicolis, G., and C. Nicolis. 1988. *Phys. Rev. A38,* 437.
Peitgen, H., and P. Richter. 1986. *The beauty of fractals.* Springer, Berlin.
Pomeau, Y., and P. Manneville. 1980. *Commun. Math. Phys. 74,* 189.
Roux, J.C., R. Simoyi, and H. Swinney. 1983. *Physica 8D,* 257.
Rössler, O. 1979. *Ann. New York Acad. Sci. 316,* 376.
Ruelle, D., and F. Takens. 1971. *Commun. Math. Phys. 20,* 167.
Schuster, H. 1984. *Deterministic chaos.* Physik-Verlag, Weinheim.
Sil'nikov, L. 1965. *Sov. Math. Dokl. 6,* 163.
Sil'nikov, L. 1970. *Math. USSR Sbor. 10,* 91.
Smale, S. 1967. *Bull. Am. Math. Soc. 73,* 747.
Sparrow, C. 1982. *The Lorenz equations.* Springer, Berlin.

Pattern Formation and Asymmetry

Auchmuty, J.F.G., and G. Nicolis. 1975. *Bull. Math. Biol. 37,* 323.
Caillois, R. 1973. *La dissymétrie.* Gallimard, Paris.
Gardner, M. 1979. *The ambidextrous universe.* Scribner's, New York.

Kondepudi, D., and G. Nelson. 1985. *Nature 314*, 438.
Kuramoto, Y. 1984. *Chemical oscillations, waves, and turbulence.* Springer, Berlin.
Luria, S. 1975. *Thirty-six lectures in biology.* MIT Press, Cambridge, Massachusetts.
Nicolis, G., and J.F.G. Auchmuty. 1974. *Proc. Nat. Acad. Sci. U.S.* 71, 2748.
Nicolis, G., and I. Prigogine. 1981. *Proc. Nat. Acad. Sci. U.S.* 78, 659.
Turing, A. 1952. *Phil. Trans. Roy. Soc.* B237, 37.

Cellular Automata

Huberman, B., and T. Hogg. 1984. *Phys. Rev. Lett.* 52, 1048.
Thomas, R. 1984. *Adv. Chem. Phys.* 55, 247.
Wolfram, S. 1983. *Rev. Mod. Phys.* 55, 601.
Wolfram, S. (ed.). 1987. *Theory and applications of cellular automata.* World Scientific, Singapore.

CHAPTER 4

Probability Theory and Stochastic Processes

Feller, W. 1959. *An introduction to probability theory and its applications.* vol. I. Wiley, New York.

Fluctuations Around Thermodynamic Equilibrium

Landau, L., and E. Lifshitz. 1980. *Statistical physics,* 3d ed. Pergamon. Oxford.

Internal Fluctuations and External Noise in Physics, Chemistry, and Biology

Gardiner, C. 1983. *Handbook of stochastic methods.* Springer, Berlin.
Haken, H. 1977. *Synergetics.* Springer, Berlin.
Horsthemke, W., and R. Lefever. 1984. *Noise-induced transitions.* Springer, Berlin.
Nicolis, G., and I. Prigogine. 1977. *Self-organization in nonequilibrium systems.* Wiley, New York.
Van Kampen, N. 1981. *Stochastic processes in physics and chemistry.* North-Holland, Amsterdam.

Entropy and Irreversibility in Markov Processes; Semigroups

Feller, W. 1966. *An introduction to probability theory and its applications.* vol. II. Wiley, New York.

Schlögl, F. 1974. In H. Haken (ed.). *Cooperative effects.* North-Holland, Amsterdam.
Schnakenberg, J. 1976. *Rev. Mod. Phys. 48,* 571.

Spatial Correlations and Critical Behavior of Fluctuations

Gardiner, C., K. McNeil, D. Walls, and I. Matheson. 1976. *J. Stat. Phys. 14,* 307
Lemarchand, H., and G. Nicolis. 1984. *J. Stat. Phys. 37,* 609.
Ma, S. 1976. *Modern theory of critical phenomena.* Benjamin, Reading, Massachusetts.
Malek Mansour, M., C. Vanden Broeck, G. Nicolis, and J.W. Turner. 1981. *Ann. Phys. 131,* 283.
Mareschal, M., and E. Kestemont. 1984. *Phys. Rev. A30,* 1158.
Nicolis, G., and M. Malek Mansour. 1984. *Phys. Rev. A29,* 2845.
Nicolis, G., and J.W. Turner. 1977. *Physica 89A,* 326.
Spohn, H. 1983. *J. Phys. A16,* 4275.
Tremblay, A.M., M. Arai, and E. Siggia. 1981. *Phys. Rev. A23,* 1451.
Walgraef, D., G. Dewel, and P. Borckmans. 1982. *Adv. Chem. Phys. 49,* 311.

Time-Dependent Behavior of the Fluctuations; Nucleation

Borgis, D., and M. Moreau. 1984. *J. Stat. Phys. 37,* 631.
Frankowicz, M., and E. Gudowska-Nowak. 1982. *Physica 116A,* 331.
Hanusse, P. 1976. Diss., Doctorat ès Sciences. University of Bordeaux I.
Nicolis, G., and I. Prigogine. 1971. *Proc. Nat. Acad. Sci. U.S. 68,* 2102.
Suzuki, M. 1981. *Adv. Chem. Phys. 46,* 195.
Van Kampen, N. 1977. *J. Stat. Phys. 17,* 71.

Effect of Inhomogeneous Fluctuations Arising from Incomplete Mixing

Kumpinsky, E., and I. Epstein. 1985. *J. Chem. Phys. 82,* 53.
Menzinger, M., and A. Giraudi. 1987. *J. Phys. Chem. 91,* 4391.
Puhl, A., and G. Nicolis. 1986. *Chem. Eng. Sci. 41,* 3111.
Puhl, A., V. Altarès and G. Nicolis. 1988. *Phys. Rev. A37,* 3039.
Roux, J.C., P. De Kepper, and J. Boissonade. 1983. *Phys. Lett. 97A,* 168.
Zwietering, Th. 1984. *Chem. Eng. Sci. 39,* 1765.

Transient Bimodality

Arimondo, E., D. Dangoisse, and L. Fronzoni. 1987. *Europh. Lett. 4,* 287.
Baras, F., G. Nicolis, M. Malek Mansour, and J.W. Turner. 1983. *J. Stat. Phys. 32,* 1.

Broggi, G., and L. Lugiato. 1984. *Phil. Trans. Roy. Soc. A313,* 425.
Frankowicz, M., and G. Nicolis. 1983. *J. Stat. Phys. 33,* 595.
Lange, W., F. Mitschke, R. Deserno, and J. Mlynek. 1985. *Phys. Rev. A32,* 1271.
Nicolis, G., F. Baras, and M. Malek Mansour. 1984. In C. Vidal and A. Pacault
  (eds.). *Nonequilibrium dynamics of chemical systems.* Springer, Berlin.
Nicolis, G., and F. Baras. 1987. *J. Stat. Phys. 48,* 1070.

Sensitivity and Selection

Kondepudi, D., and I. Prigogine. 1981. *Physica 107A,* 1.
Matkowsky, B., and E. Reiss. 1977. *SIAM J. Appl. Math. 33,* 230.
Nicolis, G., and I. Prigogine. 1981. *Proc. Nat. Acad. Sci. U.S. 78,* 659.

Molecular Chirality and Weak Interactions

Kondepudi, D., and G. Nelson. 1985. *Nature 314,* 438.
Mason, S. 1984. *Nature 311,* 19.

Information and Complexity

Chaitin, G. 1975. *Scientific American 233*(5), 47.
Chaitin, G. 1987. *Algorithmic information theory.* Cambridge University Press,
  Cambridge, England.
Eigen, M., and P. Schuster. 1979. *The hypercycle.* Springer, Berlin.
Khinchin, A. 1957. *Mathematical foundations of information theory.* Dover, New
  York.
Nicolis, G., and C. Nicolis. 1988. *Phys. Rev. A38,* 437
Nicolis, G., G. Rao, S. Rao, and C. Nicolis. 1989. In Christiansen, P., and
  R. Parmentier (eds.). *Structure, coherence and chaos in dynamical systems.*
  Manchester University Press.
Nicolis, J.S. 1986. *Dynamics of hierarchical systems.* Springer, Berlin.
Shaw, R. 1981. *Z. Naturf. 30a,* 80.
Schrödinger, E. 1945. *What is life?* Cambridge University Press, London.

CHAPTER 5

Hard Sphere Models

Bunimovitch, L., and Y. Sinai. 1980. *Commun. Math. Phys. 78,* 247.
Bunimovitch, L., and Y. Sinai. 1981. *Commun. Math. Phys. 79,* 479.
Sinai, Y. 1979. In N.S. Krylov. *Works on the foundations of statistical physics.*
  Princeton Univerisity Press, Princeton, New Jersey.

The Baker Transformation and Bernoulli Shifts

Arnold, V., and A. Avez. 1968. *Ergodic problems of classical mechanics.* Benjamin, New York.
Lebowitz, J., and O. Penrose. 1973. *Physics today,* Feb., 23.
Ornstein, D. 1975. In J. Moser (ed.). *Dynamical systems, theory and applications.* Springer, Berlin.
Penrose, O. 1970. *Foundations of statistical mechanics.* Pergamon, Oxford.
Penrose, O. 1979. *Rep. Prog. Phys.* 42, 1937.

Traditional Coarse-Graining

Ehrenfest, P., and T. Ehrenfest. 1959. *The conceptual foundations of the statistical approach in mechanics.* Cornell University Press, Ithaca, New York.
Gibbs, J.W. 1960. *Elementary principles in statistical mechanics.* Dover, New York.

Time Operator and Λ Transformation

Elskens, Y., and I. Prigogine. 1986. *Proc. Nat. Acad. Sci. U.S.* 83, 5756.
Martinez, S., and E. Tirapegui. 1984. *J. Stat. Phys.* 37, 173.
Martinez, S., and E. Tirapegui. 1985. *Phys. Lett.* 110A, 81.
Misra, B. 1978. *Proc. Nat. Acad. Sci. U.S.* 75, 1627.
Misra, B., I. Prigogine, and M. Courbage. 1979. *Physica 98A,* 1.

Light-Matter Interaction

Panofsky, W., and M. Phillips. 1956. *Classical electricity and magnetism.* Addison-Wesley, Reading, Massachusetts.
Petrosky, T., and I. Prigogine. 1988. *Physica 147A,* 439.
Prigogine, I., and T. Petrosky. 1988. *Physica 147A,* 461.

Kinetic Theory; Collision Operator

Balescu, R. 1975. *Equilibrium and nonequilibrium statistical mechanics.* Wiley, New York.
Prigogine, I. 1962. *Nonequilibrium statistical mechanics.* Wiley, New York.
Résibois, P., and M. De Leener. 1977. *Kinetic theory of dense fluids.* Wiley, New York.

Epistemological Aspects

Born, M. 1949. *Natural philosophy of cause and chance.* Clarendon, Oxford.
Prigogine, I., and I. Stengers. 1979. *La nouvelle alliance.* Gallimard, Paris.

Rosenfeld, L. 1979. *Selected papers*, R. Cohen and J. Stachel (eds.). Reidel, Dordrecht.
Sinai, Y. 1981. *L'aléatoire du non-uléatoire*. *Priroda* (3), 72. (French translation in *Ann. Fond. L. de Broglie*. *10*(4), 29 [1985]).

CHAPTER 6

Materials Science

Balakrishnan, V., and C. Bottani (eds.). 1986. *Mechanical properties and behavior of solid—plastic instabilities*. World Scientific, Singapore.
Bass, M., (ed.). 1983. *Laser materials processing*. North-Holland, Amsterdam.
Bilgram, J., and P. Böni. 1984. In G. Nicolis and F. Baras (eds.). *Chemical instabilities*. Reidel, Dordrecht.
Bottani, C., and G. Caglioti. 1982. *Physica Scripta T1*, 65.
Langer, J. 1980. *Rev. Mod. Phys. 52*, 1.
Picraux, S., and W. Choyke (eds.). 1982. *Metastable materials formation by ion implantation*. North-Holland, Amsterdam.
Walgraef, D., and E. Aifantis. 1985. *J. Appl. Phys. 58*, 688.
Walgraef, D. (ed.). 1987. *Patterns, defects and microstructures*. M. Nijhoff, Dordrecht.

Cellular Dynamics

Les défenses du corps humain. *La Recherche*. Special issue, May, 1986.
Garay, R., and R. Lefever. 1978. *J. Theor. Biol. 73*, 417.
Hiernaux, J., R. Lefever, G. Uyttenhove, and T. Boon. 1986. In G. Hoffman, J. Levy, and G. Nepom. (eds.). *Paradoxes in Immunology*. CRC Press, Boca Raton, Florida.

Modeling of Climatic Variability

Benzi, R., G. Parisi, A. Sutera, and A. Vulpiani. 1982. *Tellus 34*, 10.
Berger, A., (ed.). 1981. *Climatic variations and variability: facts and theories*. Reidel, Dordrecht.
Berger, A., et al (eds.). 1984. *Milankovitch and climate*. Reidel, Dordrecht.
Budyko, M. 1974. *Climate and life*. Academic Press, New York.
Gates, W. 1981. In A. Berger (ed.). *Climatic variations and variability: facts and theories*. Reidel, Dordrecht.
Ghil, M., R. Benzi, and G. Parisi. (eds.). 1985. *Turbulence and predictability in geophysical fluid dynamics and climate dynamics*. North-Holland, Amsterdam.
Haltiner, G. 1971. *Numerical weather prediction*. Wiley, New York.

Imbrie, J., and K. Imbrie. 1979. *Ice ages.* Enslow, Hillside, New Jersey.
Källén, E., C. Crafoord, and M. Ghil. 1979. *J. Atmos. Sci. 36,* 2292.
Lorenz, E. 1964. *Tellus 16,* 1.
Nicolis, C. 1982. *Tellus 34,* 1.
Nicolis, C. 1984. *Tellus 36A,* 1, 217.
Nicolis, C., and G. Nicolis (eds.). 1987. *Irreversible phenomena and dynamical systems analysis in geosciences.* Reidel, Dordrecht.
Saltzman, B. 1978. *Adv. Geophys. 20,* 184.
Saltzman, B. 1983. *Adv. Geophys. 25,* 173.
Schneider, S., and R. Dickinson. 1974. *Rev. Geophys. Space Phys. 12,* 447.

Social Phenomena

Allen, P. 1982. *Environment and Planning B9,* 95.
Allen, P., G. Engelen, and M. Sanglier. 1986. In *The praxis and management of complexity.* United Nations University Press, Tokyo.
Allen. P., and J. McGlade. 1986. *Can. J. Fish Aqu. Sci. 43,* 1187.
Deneubourg, J.L., J. Pasteels, and J.C. Verhaege. 1983. *J. Theor. Biol. 105,* 259.
Grassé, P. 1959. *Insectes Sociaux 6,* 127.
May, R. 1973. *Model ecosystems.* Princeton University Press, Princeton, New Jersey.
Montroll, E., and W. Badger. 1974. *Quantitative aspects of social phenomena.* Gordon and Breach, London.
Prigogine, I., and R. Herman. 1971. *Kinetic theory of vehicular traffic.* Elsevier, New York.
Prigogine, I., P. Allen, and R. Herman. 1977. In E. Laszlo and J. Bierman (eds.). *Goals for a global community.* Pergamon, New York.

APPENDIXES

Linear Stability Analysis and Bifurcation Analysis

Guckenheimer, J., and Ph. Holmes. 1983. *Nonlinear oscillations, dynamical systems, and bifurcations of vector fields.* Springer, Berlin.
Iooss, G., and D. Joseph. 1981. *Elementary stability and bifurcation theory.* Springer, Berlin.
Nicolis, G., and I. Prigogine. 1977. *Self-organization in nonequilibrium systems.* Wiley, New York.
Nicolis, G. 1981. In R. Enns, et al. (eds.). *Nonlinear problems in physics and biology.* Plenum, New York.
Sattinger, D. 1973. *Topics in stability and bifurcation theory.* Springer, Berlin.

Lyapounov Exponents

Collet, P., and J.P. Eckmann. 1980. *Iterated maps on the interval as dynamical systems.* Birkhäuser, Basel.
Lichtenberg, A., and M. Lieberman. 1983. *Regular and stochastic motion.* Springer, Berlin.
Oseledec, V. 1968. *Trans. Moscow Math. Soc. 19*, 197.
Ruelle, D. 1979. *Ann. New York Acad. Sci. 316*, 408.

Perturbation of Resonant Motions

Birkhoff, G. 1935. *Mem. Pont. Acad. Sci. Novi Lyncaei 1*, 85.
Chirikov, B. 1979. *Phys. Rep. 52*, 463.
Moser, J. 1973. *Stable and random motions in dynamical systems.* Princeton University Press, Princeton, New Jersey.

Homoclinic Points

Ekeland, I. 1984. *Le calcul, l'imprévu.* Seuil, Paris.
Newhouse, S. 1974. *Topology 13*, 9.
Newhouse, S. 1979. *Publ. Math. IHES 50*, 101.
Newhouse, S. 1980. In J. Guckenheimer, J. Moser, and S. Newhouse (eds.). *Dynamical systems.* Birkhäuser, Basel.
Nitecki, Z. 1971. *Differentiable dynamics.* MIT Press, Cambridge, Massachusetts.
Poincaré, H. 1899. *Les méthodes nouvelles de la méchanique céleste.* Gauthier-Villars, Paris.
Smale, S. 1967. *Bull. Am. Math. Soc. 73*, 747.

Reconstruction of the Dynamics from Time Series Data

Basic references

Eckmann, J.P., and D. Ruelle. 1985. *Rev. Mod. Phys. 57*, 617.
Grassberger, P., and I. Procaccia. 1983. *Phys. Rev. Lett. 50*, 346.
Packard, N., J. Crutchfield, J. Farmer, and R. Shaw. 1980. *Phys. Rev. Lett. 45*, 712.
Takens, F. 1981. In *Lecture Notes in Math.* vol. 898. Springer, Berlin.
Wolf, A., J. Swift, H. Swinney, and H. Vastano. 1985. *Physica 16D*, 285.

Applications in hydrodynamics and chemistry

Brandstäter, A., et al. 1983. *Phys. Rev. Lett. 51*, 1442.

Malraison, B., P. Atten, P. Bergé, and M. Dubois. 1983. *J. Phys. Lett. 44*, 897.
Swinney, H., and J.C. Roux. 1984. In C. Vidal and A. Pecault (eds.). *Nonequilibrium dynamics in chemical systems*. Springer, Berlin.

Applications in analysis of natural time series

Atmanspacher, H., H. Scheingraber, and W. Voge. 1988. *Phys. Rev. A37*, 1314.
Babloyantz, A., M. Salazar, and C. Nicolis. 1985. *Phys. Lett. 111A*, 152.
Babloyantz, A., and A. Destexhe. 1986. *Proc. Nat. Acad. Sci. U.S. 83*, 3513.
Berger, A., and P. Pestiaux. 1982. Tech. Rept. No. 28, Inst. Astron. and Geophys., Univ. of Louvain, Belgium.
Dvorak, I., and J. Siska. 1986. *Phys. Lett. 118A*, 63.
Essex, C., T. Lookman, and M. Nerenberg. 1987. *Nature 326*, 64.
Fraedrich, K., 1986. *J. Atmos. Sci. 43*, 419.
Mayer-Kress, G. (ed.). 1986. *Dimensions and entropies in chaotic systems*. Springer, Berlin.
Nicolis, C., and G. Nicolis. 1984. *Nature 311*, 529.
Nicolis, C., and G. Nicolis. 1986. *Proc. Nat. Acad. Sci. U.S. 83*, 536.
Rapp, P., I. Zimmerman, A. Albano, G. Deguzman, and N. Greenbaum. 1985. *Phys. Lett. 110A*, 335.
Saltzman, B. 1987. In C. Nicolis and G. Nicolis (eds.). *Irreversible phenomena and dynamical systems analysis in geosciences*. Reidel, Dordrecht.
Shackleton, N., and N. Opdyke. 1973. *Quat. Res. 3*, 39.
Tsonis, A., and J. Elsner. 1988. *Nature 333*, 545.

Primordial Irreversible Processes

General background on cosmology

Hawking, S. 1988. *A brief history of time, from the big bang to black holes*. Bantam, New York.
Weinberg, S. 1972. *Gravitation and cosmology*. Wiley, New York.
Weinberg, S. 1977. *The first three minutes of the universe*. Basic Books, New York.

Instability of the quantum vacuum

Brout, R., F. Englert, and E. Gunzig. 1978. *Ann. Phys. 115*, 78.
Brout, R., F. Englert, and P. Spindel. 1979. *Phys. Rev. Lett. 43*, 417.

Irreversible processes in early universe

Gunzig E., J. Géhéniau, and I. Prigogine. 1987. *Nature 330*, 621.
Prigogine, I., J. Géhéniau, E. Gunzig, and P. Nardone. 1988. *Proc. Nat. Acad. Sci. U.S. 85*, 7428.

Index

Printed in the United States
52822LVS00002B/13-27